The Origin of Man

GORDON W. HEWES

University of Colorado
Boulder, Colorado

Burgess Publishing Company • Minneapolis, Minnesota

A SERIES ON
BASIC CONCEPTS IN ANTHROPOLOGY
Under the Editorship of
A. J. Kelso, University of Colorado
Aram Yengoyan, University of Michigan

Contents

1, STATEMENT OF THE PROBLEM

6, EVOLUTION AND SURVIVAL

10, THE MECHANISMS OF EVOLUTION

13, VERTEBRATE EVOLUTION

15, THE FIRST PRIMATES

19, THE LATER HISTORY OF THE PRIMATES

24, EARLY APES AND EARLY MEN

32, EARLY MAN AND HUNTING

39, THE AUSTRALOPITHECINES

47, HOMO ERECTUS

50, NEANDERTHAL MAN AND OTHER
EARLY FORMS OF MODERN MAN

55, WHAT COMES NEXT?

56, REFERENCES AND SUGGESTED READINGS

# Statement of
the Problem

The problem of the origin of man has been answered or approached in many ways. In the past, man, along with the rest of the world, was believed to have been created rather suddenly, by God or some other supernatural being (Gillispie 1951; Greene 1959). The familiar Biblical account in Genesis is only one of many such creation explanations. In the last few centuries, however, explanations resting on other than religious tradition have come to the fore. Over a century ago, Charles Darwin ventured to speculate about the ultimate beginnings of life on our planet. His views are not too different from those of scientists of today who continue to speculate about this problem.

It is often said that all the conditions for the first production of a living organism are now present, which could ever have been present. But if (and oh! what a big if!) we could conceive in some warm little pond, with all sorts of ammonia and phosphoric salts, light, heat, electricity, etc., present, that a proteine compound was chemically formed ready to undergo still more complex changes, at the present day such matter would be instantly devoured or absorbed, which would not have been the case before living creatures were formed.[1]

The development of the science of geology, of the study of fossil plants and animals (Edwards 1967), and of several fields of biology has led, over the past 110 years or so, to a new view of man's place in nature, and of the origins of life and the gradual development or unfolding of life-forms which is usually called "evolution" (Tax 1960). Charles R. Darwin's *Origin of Species* (1859) is usually taken to mark this scientific revolution, although some features of his formulation had been anticipated by others (Glass et al. 1959; Chambers 1969 [1844]). Darwin's book did not deal significantly with human

---

[1]Charles Darwin. 1871. *The descent of man.*

1

origins, but several years later Darwin wrote *The Descent of Man* (1871). Although there were solid reasons for interest in Darwin's work on other grounds, the great public controversy which arose was mainly because the notion of the gradual evolution of living things obviously contradicted Biblical tradition.

It is possible to answer the question, "Where did man come from?" in a few sentences, in the light of present scientific understanding, about as follows: Mankind has descended, by a gradual process of evolutionary change, brought about by environmental pressures or "natural selection," from nonhuman animal ancestors living ten to a dozen million years ago. If these ancestors could be resurrected, they would not closely resemble any living species of ape, but we would have to describe them as "more like apes than anything else." Any such short answer is bound to be unsatisfactory, and further fails to really explain the process. In 1863, Thomas Henry Huxley was defending Darwin's views on evolution to audiences who protested that the origin of man from an apelike ancestor somehow threatened man's view of his place in the natural order. To these protests Huxley responded:

On all sides I shall hear the cry: "We are men and women, not a mere better sort of apes, a little longer in the leg, more compact in the foot, and bigger in brain than your brutal Chimpanzees and Gorillas. The power of knowledge—the conscience of good and evil—the pitiful tenderness of human affections, raise us out of all real fellowship with the brutes, however closely they may seem to approximate to us."

To this I can only reply that the exclamation would be more just, and would have my own entire sympathy, if it were only relevant. But it is not I who seek to base Man's dignity upon his great toe, or insinuate that we are lost if an Ape has a hippocampus minor. On the contrary, I have done my best to sweep away this vanity. I have endeavoured to show that no absolute structural line of demarcation wider than that between the animals which immediately succeed us in the scale can be drawn between the animal world and ourselves; and I may add the expression of my belief that the attempt to draw a psychical distinction is equally futile, and that even the highest faculties of feeling and of intellect begin to germinate in lower forms of life. At the same time, no one is more strongly convinced than I am of the vastness of the gulf between civilised man and the brutes; or is more certain that, whether from them or not, he is assuredly not of them. No one is less disposed to think lightly of the present dignity, or despairingly of the future hopes, of the only consciously intelligent denizen of this world.[2]

[2]Thomas Henry Huxley. 1863. *Evidences as to man's place in Nature.*

Moreover, such a pat answer suggests that we know much more than we really do. Specialists on the subject do not agree on all of the details although there is certainly a consensus among scientists about the major outlines. The answers are very much still in the making, and hundreds of specialists in many countries are engaged in research on different aspects of the problem. One research topic now being actively pursued is the natural behavior of nonhuman primates. John Napier, a British student of human evolution, discusses how such studies can further our understanding of human evolution:

Primate behavior field studies have vastly increased our knowledge of how the lives of primates are organized, how the environment has influenced the direction of behavioral adaptation, and has put an edge on our awareness of the part played by culture in human adaptation. Failing the direct evidence of fossil hominids, man must gain what knowledge he may by analogy—by the indirect evidence provided by the study of non-human primates. Although the hominid stem has been separated from the ape stem for many millions of years, and from the Old World monkey stem for an even longer period of time, man shares a common inheritance with both apes and monkeys—a common genetic potential. Primates, particularly anthropoid primates, are, as I have indicated above, extremely conservative animals which have departed little from the ancestral pattern of primate structure and, thus, the subsequent evolution of non-human primates has run in close parallel with that of human primates. It is a perfectly reasonable proposition to study intensively the behavior of apes and monkeys for they provide the evidence, albeit, indirect, that we lack for man. This is particularly relevant if the environments of the animals studied happen to correspond with the environment envisaged for early man. Thus ground-living primates such as baboons and macaques are probably more relevant to the human situation than the highly specialized forest-living chimpanzees and gorillas. Perhaps even more relevant are field studies of primates in transition from an arboreal mode to a terrestrial one. The savanna monkeys of Africa (Cercopithecus aethiops) *provide the best example of this category. Savanna monkeys provide an evolutionary model for the metamorphosis of a species transferring from a forest to a savanna habitat, an ecological progression that, as we see it at present, re-creates a critical phase in the story of our past.*

The real significance of the study of monkeys and apes is that non-human primates provide a mirror into which man may look to discover what being a man is all about. But let us look at our reflected image critically and not emotionally; empirically and not intuitively; with reason and not with prejudice.[3]

[3] John Napier. 1970. *The roots of mankind: the story of man and his ancestors.* London: George Allen & Unwin, pp.218-21.

3

The name for the process or problem, by the way, is *hominization* (becoming human), or *anthropogenesis* (origin of man), and it is worth stating at this point that the process has by no means come to an end; man seems, like other living things, to be still evolving. As long as we live in a changing world, we will continue to evolve, along with all the other plants and animals on this planet.

We must hasten to add that "evolution" does not necessarily mean "progress." The Latin word from which our English word comes means to "unfold" or "unwind," which does suggest growth. However, if progress means movement toward a definable goal, we may be misled; scientific evidence for any such goal has not been forthcoming. It is true, nevertheless, that life on this planet has become more complex in the course of time. Meanwhile, as representatives of the most intelligent form of life to emerge so far on this planet, we should find it worthwhile to examine the record and see how we have arrived at this level of development.

The kinds of evidence we have about human origins include the quite solid pieces of fossil bone and teeth—rather few and fragmentary—far back in the geological time span during which hominization has been taking place but becoming more complete as we approach recent times. These fossils can be dated with sufficient precision in enough cases so that the specimens we cannot directly date can usually be fitted into the sequence with some confidence. The dating techniques, chiefly based on rates of radioactive decay of certain elements, have only been available for the last twenty-five years; before that time there was strong disagreement as to the age of certain hominid fossil remains, and a general inclination to compress human evolution into the last million years or so. New dating techniques have had the result of spreading out the time during which human evolution was going on. In addition to fossil remains and the datings associated with them, we can utilize a very large part of the entire field of biology, including data on other forms of life—some of them far distant from man and the other mammals. However, certain biochemical substances occur in all known living things, strongly suggesting a common origin for all life—plant, animal, and at a microscopic level where the distinctions between plant and animal cease to be significant. The basic uniformities of living things at the molecular level are striking. John Buettner-Janusch of Duke University discusses the importance of amino acid sequences in protein molecules for genetics and evolution of living organisms, and for the evolution of primates in particular:

The amount of genetic and evolutionary information contained in a sequence of amino acids is enormous. . . . We reemphasize that a study of the sequence of amino acids in a protein produces some of the best genetic data available today. If we can analyze a single protein in its homologous forms in a variety

4

of related organisms, we can determine indirectly the kinds and magnitude of genetic changes which took place during the period since the organisms concerned branched from a common stock.

At the molecular level, there are a number of logically distinct approaches to a study of evolution. We can sort most of the theoretical approaches, the discussions, and, indeed, the polemics into two groups. First, there are the theories or approaches which take for granted, or insist, that molecular evolution can be studied independently of organisms. A basic assumption is that molecules evolve as such and that we can reconstruct this evolution. The question asked in this approach is: Can classifications of the Primates be made or unmade on the basis of the present studies of protein molecules? Enthusiastic proponents of this view are convinced that phylogenies of organisms, even classifications, may be made on the basis of molecular events. They imply that disagreements about taxonomy can be resolved with the molecular approach.

More cautious is the second view that molecules evolve as parts of organisms, indeed as parts of populations of organisms. There is no question that we can analyze the evolutionary changes in molecules themselves without referring in detail to the organisms, but we must remember that molecules evolve as parts of complex systems. Therefore we must ask whether molecular evolution can be observed as part of an accepted phylogeny of organisms. Can we correlate the evolutionary changes of hemoglobin with an accepted or acceptable classification of the Primates?[4]

[4]John Buettner-Janusch. 1966. *Origins of man: physical anthropology.* New York: John Wiley & Sons, pp. 555-7.

Evolution and Survival

The body of scientific knowledge known as biology has been built up mainly within the past 150 years, and comprises not only the fossil or paleontological record, but information on anatomy or structure, physiology or life-processes, and behavior, growth, reproduction, and environmental and distributional information as well. Human evolution is a subject which has not been pursued in isolation, but has been seen ever since evolutionary biology appeared probable, as only a part of the living world of plants and animals. In genetics, for example, much of our knowledge comes from work with plants, or animals such as fruit flies, guinea pigs, or mice. Much of the science of psychology rests on studies of the behavior of nonhuman animals such as rats, dogs, or cats. Although the importance of human evolution may loom large because we happen to be human beings, it would in principle be possible to comprehend most of the important processes which have led to the appearance of mankind without ever studying human beings or their remains.

We shall not try here to present the details about genetics, or about the basic molecular biology of the living cell (Alland 1969; Cain 1960; Dobzhansky 1955; Dunn 1969; Kelly 1966). Good surveys of these topics can be found in modern biology texts.

In the evolution of biosexual species (of plants or animals) it is critical that males and females, each carriers of specialized kinds of reproductive cells, get together or mate so that their genetic material may be recombined or reshuffled, which results in successive rearrangements of genes or units of heredity. This elaborate process has evidently been developed more than once in the history of living organisms, to promote sufficient variability or flexibility in each generation so that the survival of the species has a better chance than if all offspring in each generation were absolutely alike in their genetic constitution or outward form. Since this statement sounds too purposeful, it would be an improvement to add that over time the existence of two sexes, engaged in seemingly endless reshufflings of their genes at each mating, has proved to be effective in increasing the survival chances for plant and animal species, particularly when they consist of individuals having very large numbers of body cells and complex body parts. The "sexes" may not have to exist in separate individuals; in many plants and some animals, both "male" and "female" sex cells and organs are present in the same individual, but even this combined arrangement still serves to maintain an effective level of genetic reshuffling.

It is hard to avoid the notion that, given the elaborate biological mechanisms which have evolved to "ensure" greater survival, survival itself is not some kind of goal or purpose, or some kind of game in which living forms are in a perpetual contest with nature to see whether particular species or lineages will "win" by resisting extinction. However, such a view is a form of vitalism, which attributes some kind of metaphysical propensity to survive on the part of certain complex molecular systems. A harder-headed, more materialistic view is that just as masses of matter cannot help conforming to the laws of motion, or light waves to the properties of light, and so on, organic systems, once having come into existence, cannot exist without exhibiting processes or behaviors which tend to their perpetuation under particular conditions. The "will to survive" may provide food for philosophical or religious contemplation, but it is a will-o'-the-wisp as far as science is concerned; our human tendency to think in such terms presumably arises because we alone of all plants and animals have arrived at an awareness of individual death, and an ability to think about the survival or extinction of a whole species.

One of the mechanisms which increases the possibility for species survival is of course production of a large number of offspring, so that even if most of them are eliminated before they in turn can reproduce the next generation, enough will survive to ensure that the overall population size of the species will not dwindle. Thus, a great many forms of life exhibit a truly profligate "overproduction" of new individuals, all but a few of whom die before they in turn can similarly reproduce. An alternative to this lavish reproductiveness is to concentrate such care and protection on a small number of offspring that relatively few will fail to reproduce in turn. No plants, and only a few kinds of animals have ever moved very far in this direction. The reproductive potentials of plant and animal species are vastly greater than any actuality, thanks to the limited availability of food, space, or energy which in the case of the higher, photosynthetic plants, can be obtained directly from sunlight. The fundamental insight of Darwin and Wallace, in the 1850s, came from this realization about the checks and balances in nature which keep any one species of plant or animal from running out of food or space at the expense of all other forms of life. True, this ecological balance is precarious and, in the view of many specialists today, man in particular poses such a threat. There is nothing which *ensures* the continuation of diverse forms of life on a planet such as ours, or which can be depended on to prevent a semi-intelligent bungler like *Homo sapiens* from wrecking the whole system. Two or three hundred years ago, scholars were more optimistic; there was a widespread belief in the "Great Chain of Being," a theory which held that the Creator had separately fashioned each species to perfection, each in its proper place in nature (Lovejoy 1936). In the usual version of this view, it was held that no species created by God could ever become extinct (except perhaps at the End of the World), since that would imply some mistake on the part of the Creator. Instead, paleontology, has revealed that extinction is the fate of most species, and that very few forms of life have maintained themselves in essentially unchanged form for more than a fraction of the time life has existed on earth.

When we speak of the *origin of a species* like modern man, we mean that a line of organisms, related through successive generations, has changed, normally by slow and small degrees, from a species we would describe under one name or *taxon*, into a species we would call something else (Cain 1960). In most cases, this would involve some obvious outward changes—that is, among both plants and animals, members of this or that recognized species actually look different from members of other species. There are likely to be visible differences in size, shape, color, etc., beyond the kinds of differences we may be able to detect between individuals within the same species. To be sure, we may run into complications. Many plants and animals look very different at different stages of their life cycle; the caterpillar may not resemble the adult moth. The sexes may be highly differentiated, and so on. A less obvious difference between closely related species is that they may not be able to breed with one another, or if they can, few or none of their offspring can in turn produce offspring. Perhaps the commonest example of this is that horses and donkeys, clearly similar in general appearance, can breed and produce offspring we call mules, but mules are sterile. On the other hand, some clearly separate species, such as lions and tigers, which normally breed "true to type"—that is, for generation after generation their successive offspring look like lions or tigers—can, under the abnormal conditions of captivity, breed and produce offspring we call *ligers* or *tiglons*. Domestic dogs can mate with wolves, producing fertile offspring. There are species which can produce fertile offspring only within their own species (man may well be such a case) and others which, though now members of different species according to accepted zoological criteria, can interbreed and yield fertile offspring. There seems to be a long-term advantage, however, for species to develop barriers to cross-breeding. Various mechanisms exist to ensure that mistakes in mating do not often occur, such as peculiar color patterns, species-specific mating songs (in birds), distinctive scents, elaborate mating rituals, or biochemical incompatibilities. We must be on guard, here as elsewhere when considering biological processes, to avoid concluding that this means that there is some general "natural law" which somehow insists on purity of lineage. Barriers to cross-specific mating are an evolutionary response to the existence of biochemical and other diversities which would render cross-specific matings a waste of effort, and confine sexual encounters to those which are more likely to result in viable offspring. The original causes of biological incompatibility may have been as neutral as the formation of a mountain or water barrier, or climatic differences. When physical barriers exist for a long time, it seems also certain that species separation will occur, and genetic differences will accumulate until two populations have diverged from a single ancestral stock. If one or both of the separated populations happened to be very small, "genetic drift" and "founder effect" might be the cause of divergence, but the chief factors in most cases of such geographical separation seem to come from significant differences in living conditions on the two sides of the barrier. Needless to say, we have chosen a two-sided barrier merely for the sake of illustration. Plants and animals are forever moving not just into two significantly different environments, but three, four or more at the same time.

The Mechanisms of Evolution

The prime movers of biological differentiation seem to have been mainly of the geographical environmental type, plus the fact that all life on the planet is subject to low-level amounts of radiation either from outer space as cosmic rays, or from the rocks and soils (Dobzhansky 1955). While all living things have evolved so as to resist such background radiation, the genetic constituents of individual cells are subject to destruction or injury capable of producing abnormal cell growth, or in the case of reproductive cells, hereditary changes known as "mutations." Otherwise, the reproductive cells (which develop into sperm cells or ova) are fairly well insulated from external damage capable of altering their genetic codes. Various chemical substances also may occur in the environment, either naturally, or, increasingly, as a result of human technology, which may affect the genes; but for most of them plants and animals have developed various ways of detecting and coping, by means of taste and smell, or by means of internal mechanisms for neutralizing or rejecting potentially dangerous materials.

Darwin and his contemporaries were aware of the seemingly universal tendency for living things to exhibit small variations, within species, and it was on the basis that small variations existed that he and Wallace developed their idea that nature acts as a selective agency, favoring certain variations under certain environmental conditions, and reducing the survivorship of unfavorable variations under the same conditions (deBeer 1965). For Darwin these variabilities were simply a "given," for which no real explanation was provided in the initial formulation of the theory of evolution through "natural selection." "Natural selection," by the way, is simply a phrase like "artificial or man-directed selection or breeding," as with plant or animal breeding, except that nature is assumed to select without any long-range goal or purpose. It proved quite difficult for many people to accept the idea that the rich diversity of living things, seemingly so ingeniously adapted to particular ways of life or niches in the environment, had gotten that way by what was basically only a random or chance process. The term "survival of the fittest," coined not by Darwin but by the sociologist Herbert Spencer, helps to explain how a random process could result in what seems, in the long run, to look purposeful, but it also conjures up a kind of tooth and claw competitive struggle which, while present in some of the animal kingdom, is not the most characteristic kind of biological competition. Plants, for example, do not physically fight each other for food or living space, and even where tooth and claw struggles can be observed, as with prey and predator animals, the "winner" who ends up eating his prey cannot be considered to have done so in order to eliminate one more "competitor"; he has merely obtained some food, and if he were 100 per cent successful in killing his "competition" he would have eliminated his own food supply.

It was not until the early twentieth century that experiments were able to detect that one source of "inherent" variability in organisms came from impinging radiant energy—with fruit flies as subjects. More recent experiments have shown that low-level radiation in controlled amounts is able to induce mutations in plants and insects, which can then be followed in subsequent generations of breeding experiments.

As in other animals and plants, our remote ancestors must have exhibited a range of small variability in each generation, which was played upon or selected from, as a consequence of changing living conditions. As divergences became more marked, barriers to mating must also have developed, as they have in many other species—because such mating incompatibilities, too, proved to be advantageous for survival. "Advantageous for survival" is really a circular notion; we only make this judgment by hindsight. Whatever enhances the chances of one group of individuals within a population to produce more of their own descendants in the next generation is advantageous. However, things do not go on "ad infinitum," and conditions change, so that what was once advantageous becomes a liability, and vice versa. Whereas it must have been worthwhile for man's earlier ancestors, along with the rest of the primates, to adapt more and more successfully to life in the trees, there came a time when this ceased to pay off, and it became more favorable for survival for the population ancestral to man to live on the ground. It is possible that the features we consider "racial" were adaptive or biologically useful for some time in man's past (Coon 1963, 1965), but it is also likely that now and in the future, the existence of marked regional or geographical differences in human gene frequencies could be more damaging to man's long-term survival than the development of a more homogeneous human species.

The story of the vigorous controversy which arose after Darwin's publication of *Origin of Species* in 1859 need not be retold in detail (deBeer 1965). Much opposition came from outraged theologians and other support-

11

ers of a literal Biblical interpretation of Creation. But there was also argument within the evolutionist camp. Darwin believed that natural se'ection operated to pick out the small variations which occur naturally in any living species. Some of his closest supporters, like Thomas Huxley (1863[1921]) and Francis Galton (his cousin), favored the notion that evolution proceeded by sudden leaps or "sports" as plant and animal breeders referred to unexpected outcomes of crossing different lines. We referred to them earlier as *mutations*. For years this argument went on, and it was not resolved by the rediscovery around 1900 of Mendel's genetic principles. Many of the students of the rapidly expanding science of genetics argued that evolution was a matter of mutations only, rather than of natural selection. On the other hand, a group of biologists who developed the statistical techniques then known as biometrics tended to play down the evolutionary importance of mutation. This bitter argument did not die down until the 1920s, when a new, broad, "synthetic" approach to evolutionary biology emerged from the work of biologists like Haldane, Sewall Wright, and others. Their new models of evolution resolved the seeming contradictions between the views of the "mutationists" and the "biometricians." In succeeding decades, further insights were opened up through the development of molecular biology, including the "decipherment" of the DNA genetic code.

There is no longer *a* single "theory of evolution," to be arrayed against a traditional Creationist doctrine, but rather it can be truthfully said that evolution permeates almost all aspects of modern biological science, making it practically impossible to understand the living world except as the outcome of continuous and continuing evolutionary processes. Even the geological history of the earth's crust, and of the atmosphere around this planet, over the past half-billion years, cannot be understood without assuming a gradual emergence of various, interrelated forms of life.

There is as yet no firm agreement on a model for how living systems might have arisen on this planet out of inorganic molecules, although there are some promising theories. Long after the first tiny unicellular living systems appeared, the next major biological event occurred—the evolution of plant-organisms containing a pigment (chlorophyll) enabling them to utilize the energy of sunlight to convert mostly water and air (plus minute traces of metallic and other elements) into live tissue, chiefly turning carbon dioxide into sugar. This breakdown led to the gradual oxygenation of the atmosphere, permitting a great increase in the capacity of the sunlit waters and soil-surfaces of the planet to support living matter, and the emergence of higher animals which, directly or indirectly, depend absolutely on the power of green plants to utilize sunlight.

Vertebrate Evolution

We cannot stop to trace the long and immensely complicated history of life from its unicellular nonphotosynthetic beginnings, but must move far ahead to pick up the story of a very limited set of animals which had evolved from early reptiles (Romer 1971; Time–Life Books 1972). This group of animals, like primitive reptiles, had backbones, four limbs, and other basic anatomical features inherited by the reptiles from still more ancient amphibian ancestors, which in turn stem from primitive fishes. But, unlike ordinary reptiles, the most primitive mammals had added some critically important features: warm-bloodedness, which means that the body maintains a fairly uniform temperature despite fluctuation in the outside temperature; specialized insulation, derived from reptilian scales, but now in the form of hairs; and provision of a special infant food—milk. Another feature of mammals is that most of them possess specialized, highly differentiated teeth, which come in two separate "sets," whereas reptiles have simpler, peglike teeth which break off and fall out and can be replaced by new ones. These specialized teeth permit mammals to handle a variety of foods neglected by reptiles (such as small hard seeds, or grass), or, in the case of a meat diet, to chew it up thoroughly and thus greatly accelerate the subsequent processes of digestion. Some of the other features we usually associate with mammals were already present among certain reptiles, such as live birth rather than laying eggs, and a four-chambered heart, found among crocodilians. Birds are also warm-blooded (and in a less efficient fashion, so are some fast-swimming fishes, such as the tunas), have four-chambered hearts and effective special insulation (feathers) derived from scales, but it can be shown that the birds are a separate evolutionary stem from a later group of reptiles (the dinosaurians). The fact that both mammals and birds became prominent toward the close of the great Age of Reptiles or Mesozoic (even though primitive mammals or mammal-like reptilians had been around during the entire Mesozoic as a very minor feature of the animal scene) suggests that temperature played an important part in the extinction of the larger reptiles, and the confinement of most large forms of reptilian life thereafter to tropical regions. Birds and mammals, with their insulated, warmer, thermostatically regulated bodies, could survive and remain active under temperature conditions which render reptiles sluggish or practically immobile.

The First
Primates

During the final geological subdivision of the Mesozoic, called the Cretaceous, although dinosaurs were still abundant, several kinds of mammals were present, as well as primitive birds. It was in the Cretaceous that the first primate or primatelike mammal appeared: *Purgatorius,* from Montana. As mentioned, there were by this time several kinds of mammals—a few archaic egg-layers like the surviving echidna and platypus, larger numbers of species of marsupials or pouched mammals, and among the placental (with an "afterbirth") mammals, the insectivores and primitive primates. The familiar grass-eating herd mammals of later times were conspicuously absent, since the grasslands had not yet developed. In fact, much of the terrestrial plant life was different from the present. The great proliferation of the higher flowering plants had not yet taken place. It is also well to bear in mind that the familiar shapes and locations of the continents did not then exist. Geological research in the last decade and a half shows that most of the earth's landmasses had long been fused into a single super-continent, "Pangaea," which had only started breaking up and drifting apart around this time (Time-Life Books 1972). When the continents formed a nearly continuous great landmass, animal and plant life on the land could spread fairly uniformly. When ocean gaps appeared as the landmasses rifted apart, water barriers were formed which led to evolutionary differentiation. These phenomena help to explain why Australia and New Guinea today hold most of the world's different kinds of pouched or marsupial mammals, and why the large island of Madagascar, off southern Africa, has none of the commoner large native mammals of Africa, but instead conserves a group of primates—the lemurs— who have either become extinct elsewhere, as in the Americas, or rather unimportant in Africa and Asia. Many of the formerly very puzzling problems of the distribution of fossil animals and plants now make sense in the light of the new *plate tectonics* view of continental drifting and re-collision. Elwyn Simons, a Yale University specialist on early primate fossils, suggests that continental drift explains some otherwise puzzling geographical distributions of fossil remains of early Eocene epoch age (60 to 70 million years ago) in Europe and North America:

Recent studies by McKenna and Szalay (personal communication) indicate that continental drift played an important role in the early history of the primates, because there is now impressive evidence that the final rifting of the North Atlantic between Greenland and Spitzbergen probably did not come until mid-Eocene times and that the land bridge there then lay in more southern latitudes than these areas occupy today. This discovery would explain the phenomenon originally noted by Cope (1872) that the early Eocene faunas of England found in the London clay contain many of the same mammal (and other vertebrate) genera that occur in the North American Wasatch, and in some cases both have probably identical species. The provincial ages concerned are the Wasatchian in North America and the Sparnacian in Europe. . . . We know now that in Wasatchian-Sparnacian times several primate genera, Plesiadapis, Phenacolemur, *and* Pelycodus, *occurred in both continents. Other pairs of genera such as* Teilhardina *from Belgium and* Loveina *from the Western states are closely similar.*[5]

In addition to a strikingly different world map of landmasses and oceans at the time primates began to appear, a closer look at the surface would have shown that far more of it was warm if not tropical. Although there were deserts and semi-deserts, there were also much greater areas of forest than today containing familiar-looking coniferous trees, but few of the broad-leafed and deciduous trees so conspicuous in much of the remaining forested cover today. The most obvious difference in the Cretaceous vegetation-map, and for a long time thereafter, would have been the absence of extensive grasslands. Not only were there no great herds of grass-eating mammals, but the now very abundant rodents were also absent, along with the many kinds of small, seed-eating birds.

The primates of this time, and on into the first periods (Paleocene and Eocene) of the ensuing Cenozoic or so-called Age of Mammals, would have looked more like modern rats and squirrels than monkeys, much less apes. Indeed, it is entirely likely that the earliest rodents evolved *from* primitive primates.

The general term for these primitive primates of seventy-odd million years ago, and for their surviving closest descendants, the lemurs, tarsiers, and lorises, is *prosimians*. They were tree-dwellers (some may have taken to the

[5] Elwyn L. Simons. 1972. *Primate evolution: An introduction to man's place in nature*, New York: The Macmillan Company, pp.19-20.

ground in a rough approximation of the life-ways of some rodents), equipped with good eyesight and hands and feet with padded but not clawed digits. A close examination of their jaws would have revealed a full set of specialized teeth for cutting (incisors), piercing (canines), and tearing and grinding (molars and premolars). Most of the prosimians lived on a mixed diet of insects, small seeds and fruits, and an occasional bird's egg. Whatever happened to exterminate all of the giant reptiles, and pave the way for the predominance in the past 65 million years of mammals and birds remains unclear. The mammals themselves do not appear to have suddenly acquired a fatal superiority over reptiles simply by attacking their somewhat vulnerable eggs. For one thing many reptiles bore their young alive (as do many snakes today). The big reptiles also disappeared from the sea, save for the croco-dilians and turtles, and mammals cannot be credited with this either; sea-going mammals did not evolve until well after the end of the Age of Reptiles (Romer 1954). A general cooling could have been the major cause of great reptile extinction; the more cold-adapted mammals and birds were able to populate the colder zones of the landmasses, where they underwent active evolutionary development. Aside from the possible emergence of the important Order of Rodents from some prosimian stock, the early primates, to be frank, do not seem to have played a very dramatic role in the general transformation of terrestrial animal life in the Mesozoic; the big changes came with the rise of mammalian forms with tough hoofs, specialized grinding teeth to handle grass or shoots, leaves, and buds of other tough plants, and elongated and otherwise specialized digestive tracts able to extract a nutritious diet from unpromising quantities of low-grade plant roughage. The formation of various kinds of grazing and browsing mammals, many of which tended to move about in large herds, also promoted the evolution of specialized flesh-eaters who preyed on them—notably members of the dog, cat, and bear families. Significantly, neither the big grass-eating hoofed mammals nor their carnivorous predators reached the detached Australia-New Guinea landmass, nor even the chunk of east Africa known as Madagascar. For a long time, moreover, South America, which had drifted west from Africa, was the home of a strange assortment of mammals including giant sloths and giant armadillos, and remained inaccessible to the emergent hoofed beasts evolving in the interconnected landmasses of Eurasia and Africa. Eventually South America was linked to the north by a land bridge at Panama, and was forthwith invaded by hoofed mammals of both the even- and odd-toed lineages.

The prosimian primates were in fact by this time rather conservative mammals, clinging both literally and figuratively to the forests, which were giving ground in many areas to other kinds of vegetation affording less protection to confirmed climbers, but in fact far more food to creatures on the hoof, whose teeth, stomach, and intestines could cope with browsing or grazing. There is fossil evidence for the enlargement of the world's grasslands, and for the evolutionary proliferation of new species of grasses. Ecological change affects both plant and animal life. Primates continued to live in the trees, and the venture of some prosimian primates into life on the ground, as specialized gnawers, was a failure. Eventually this niche in the environment was occupied by various rodents and by the Lagomorpha (hares and rabbits) plus their small predators such as weasels. Skeletal remains, consisting mainly of teeth and jaws, of fossil prosimians have been found in some abundance in the Rocky Mountain region of North America, in France, and in China, from the Paleocene and Eocene, and for the next period, the Oligocene, also in Egypt, in the Fayum Depression west of the Nile. Africa, evidently long involved in primate evolution, since the prosimians were among the few successful mammalian colonizers of the Madagascar landmass which rifted off from southeastern Africa, is otherwise remarkably poor in discoveries of early fossil primates.

The Later History
of the Primates

By the Oligocene Period, monkeys were present, with fewer teeth than their prosimian ancestors and less foxlike narrow snouts, bigger and more complex brains, and a reduced sense of smell (LeGros Clark 1959; Pilbeam 1972; Simons 1972). These early monkeys were still a conservative lot, as placental mammals go. The primates retained their ancient five-digit limbs, with simple fingernails instead of claws or hoofs, did not develop horns, and kept a generalized digestive system. The monkeys, moreover, had restricted numbers of offspring to one at a time; these were still fairly helpless at birth, though at least not blind like the cats and canines. In North America, where prosimians had been abundant, they became extinct, but not before some of them had evolved into an independent branch of monkeys, the Platyrrhine or New World monkeys. Another, more recent view is that the New World monkeys did not descend from the now extinct North American prosimians, but reached South America by "rafting" on floating trees at a time when continental drift had only separated that continent by a narrow channel from the western coast of Africa. These forms lost fewer teeth out of the original large (44) number of the primitive insectivores, had differently placed nostrils (the term "platyrrhine" means flat-nosed, which hardly describes the New World monkey nose) and some other distinctive traits. Many, but not all, have prehensile or grasping tails, not present in any Old World monkeys. While some Old World monkeys took to the ground (such as the ancestors of the baboons and the patas monkey, and, to some extent, the macaques), none of the New World monkeys ever forsook the trees. Today the New World primates extend from southern Mexico into the tropical forest regions of South America, in forms such as the woolly monkey, spider monkey, howler monkey, uakari, capuchin (the familiar organ-grinder's friend), and the small, spectacularly furred marmosets.

The forest way of life followed by the majority of primates was not a complete biological dead end (Dolhinow 1972; Eimerl and DeVore 1965; Jolly

1972). It was relatively safe from predators. Only some tree-dwelling snakes, occasional big predatory birds like eagles, or tree-climbing cats menace monkeys, and monkeys can evade most of these dangers by remaining highly alert for forms that look snakelike or for swooping birds; by climbing higher, up into the uppermost foliage or by leaping to another branch or tree they can usually escape the cats.

More important, tree-dwelling monkeys live in an environment which lays a heavy stress on acrobatic skill, coordination of body-movements, and vision. They see in colors, and in three dimensions. Smell, however, is of diminished importance in tree-dwelling forms (as it is in birds); aside from the prosimians, primates are inept at tracking a scent, and would not be warned by their sense of smell of some approaching carnivore. The higher primates use smell to check out their food, and in sexual contacts, but for little else.

Although wild primates today are mostly confined to the tropics and subtropics, and there chiefly to rain-forests, (there are a few important species of primates which inhabit drier tree-savannas and rocky highlands) they formerly extended into Europe as far north as Britain, and into many areas of Africa and the Middle East now too dry to support even baboons or patas monkeys. A macaquelike monkey survived until the Pleistocene in Greece, and one tiny colony of semi-wild monkeys still clings to the Rock of Gibraltar, at the southernmost tip of Spain. Although the Gibraltar monkeys have been replenished by man in recent decades, it seems that the same species once was fairly common in southern Spain in prehistoric times. Across the straits, in Morocco and Algeria, such macaques, known as Barbary "apes," inhabit pine forests and thick brushy uplands, although they are separated by thousands of miles of desert from their closest macaque relatives in the Indian region.

A huge former forest zone, from western Africa and western Europe across to India, Southeast Asia, into Indonesia, and north into China and Japan, harbored a wide variety of Old World monkeys. Drying climates have broken this formerly nearly continuous zone into islands of forest, moister grassland-savanna and deserts, and further interruptions have been brought about since the end of the last glacial period by the rise of world sea-levels, transforming Indonesia from a great southern peninsula of southeast Asia into an archipelago. At any rate, the monkeys did not penetrate beyond the eastern islands of Indonesia, and even there some had crossed the famous zoogeographic line first noted by Alfred Russel Wallace and named in his honor which separates Borneo from the Island of Celebes (Sulawesi). Man was the first primate to enter the New Guinea-Australian region, perhaps 40,000 or fewer years ago. Monkeys did reach the Japanese islands, probably at one of the times when they were joined to the Asian mainland, and it is there that we find the northernmost primates other than man—surviving under conditions of cold, snowy winters at the north end of the main island of Honshu. We have spoken of Old World "monkeys"—a very broad term for all of the Catarrhine forms, which includes a probably rather old split between those which, like ourselves and the apes, have simple, generalized digestive systems, and a group now specialized to live on leaves (the colobines) which has required the development of a more intricate, sacculated stomach and longer intestine. It seems fairly clear that the primates we call "apes" in English (the Germans and French use the same words—*Affen* and *Singes* respectively—for both monkeys and apes) are rather early offshoots probably at two different times, from the more generalized Cercopithecine branch of the Old World monkey stem. "Apes" differ from "monkeys" in several features: complete loss of the external tail (although some Old World monkeys also lack an external tail), larger body size, and, most important, differences in front and hind limb proportions, with broader shoulder-girdles conducive to an arm-swinging mode of moving through the trees (brachiation). Except for the gibbons, the anthropoid or manlike apes also have wider pelves than monkeys. We obtain a restricted view of the apes if we limit ourselves to the few surviving groups of gorilla, chimpanzee, orangutan and gibbon, and their confinement to a small portion of the remaining tropical rain-forest zone of the Old World. Several million years ago there were many more kinds of apes all the way from southwestern Europe and Africa to Indonesia, Southeast Asia, and South China. These included subgenera and species of the genus *Dryopithecus*, which earlier investigators split into many separate genera (cf. Pilbeam 1972, and Simons 1972): *Sugrivapithecus, Sivapithecus, Proconsul, Ramapithecus*, and *Gigantopithecus*—the last being a very large ape found in South China and India. *Ramapithecus*, represented by remains found both in India and East Africa has been claimed by some as a direct ancestor of mankind, chiefly on the basis of dental and jaw features. The complex arguments about *Ramapithecus*, as well as the general problem of what to do about the other mid-Cenozoic ape fossils, may be followed in the books by Pilbeam and Simons noted above. *Oreopithecus* (which means "mountain ape"), found in Europe, appears to have been very divergent from the above fossil apes.

For some of these forms the data are fairly complete. One of the *Oreopithecus* finds is a nearly complete skeleton, from an Italian coal mine deposit. *Proconsul* is represented by a crushed but otherwise nearly complete skull, and there are some longbones from *Dryopithecus* as well as from *Ramapithecus*. The rest, however, are known to us from teeth and jaw fragments, which survive better than the other parts of the skeleton simply because teeth are not only the hardest items of the skeleton, most resistant to the destructive action of soil chemicals, but also because carnivores and scavengers tend to leave sharp, indigestible teeth and accompanying jaws alone.

Monkeys and apes did not penetrate the northerly zone of Eurasia, but otherwise seem to have found habitable most of Africa, southern Europe, southern and southeast Asia, and the area of what is now Indonesia, as well as China (only monkeys, not apes reached the Japanese Islands). These regions were not exactly geologically quiescent during the Cenozoic Epoch. The formation of the great Tibetan Plateau and its southern mountain chain, the Himalaya, was mostly a Cenozoic event. From Israel and the Red Sea to the Zambezi, a huge system of rifts or crustal cracks appeared, flanked by active volcanoes, known as the Great Rift Valley. This is the beginning of a splitting off of a tectonic plate from the main body of Africa (Bishop and Clark 1967). Many very important discoveries of fossil apes happen to have been made along the central part of the eastern branch of the Great Rift Valley in Kenya and Tanzania. Although the processes of rifting and vulcanism were gradual, they produced at least local or regional changes in climate and vegetation, forming and later draining lakes, and, by thrusting up of volcanic cones, creating "rain shadows" on their lee sides. The uplift of the Tibetan Plateau likewise produced widespread climate (and hence vegetational) shifts in Central Asia, by blocking most of the warm moist air masses from the tropical Indian Ocean which would otherwise have carried rain into the heart of Asia. These geological changes would have in themselves been sufficient to account for environmental pressures leading to evolutionary change. But it must be remembered that because of the fantastically interrelated character of the living world, no important evolutionary changes in one significant species or group of species, plant, animal, or viral, can avoid affecting nearly all the rest in the long run. The rise of new species of insect-eating birds obviously imposes new selective pressures on insects, but this also affects mammals dependent on eating insects, plants which are either eaten by, or pollinated by, insects, and so on and on, in ever-widening circles of biological influence and feedback.

Hominid emergence seems to have been triggered by major climatic changes. David Pilbeam, a Yale paleontologist, writes:

The hominids appear to have evolved as forest or forest-fringe animals feeding in open spaces in wooded areas. During the Pliocene a general cooling of Earth's atmosphere combined with increasingly seasonal climates seems to have reduced the amount of evergreen forest and replaced it in many areas by deciduous woodland and tree savanna.[6]

[6]David Pilbeam. 1972. *the ascent of man: an introduction to human evolution.* New York : The Macmillan Series in Physical Anthropology, pp. 152-3.

In speculating about the kinds of landscapes in which early apelike forms might have been encouraged to come to the ground and begin evolving in a manlike direction, we should not confine our thinking to the woodland savannas of East Africa where the earliest remains of upright-walking manlike apes or apelike men have been recovered. The shoreline or littoral zone has been a favorable environment for various Old World monkeys, providing tree shelter, and both plant and easily collected animal food inhabiting the beach-line, rocks, and tidal pools. We have not found remains of early manlike forms in such environments, but that is not proof that they avoided the coasts, but only that such shorelines are highly perishable geological features, liable to be eroded or submerged (although they can also at times be uplifted and exposed well above sea-level). A few authors have been impressed by the favorable conditions that a coastal, beachcombing way of life would have had for early hominid primates.

Early Apes
and Early Men

By "Hominid" we refer to the family *Hominidae* in the zoological classification system, which comprises modern mankind: *Homo sapiens*, and all of the fossil forms which can be recognized as already distinctively manlike rather than like the three anthropoid apes—gorilla, chimpanzee, and orangutan—which form the family *Pongidae*, along with fossil remains related to them (Washburn 1963). The gibbon of Southeast Asia and Indoñesia is seemingly the product of an earlier and separate branch of anthropoid ape evolution, and forms another Family, *Hylobatidae*, which may have split off as far back as the Oligocene. Although the data (as stated, chiefly dental) are not absolutely clear, it seems probable that the ancestors of the living Pongidae and the Hominidae formed a common stock well into the Miocene—30 million years ago or so—and that a definite split leading to the final separation of pongids and hominids came either late in the Miocene or in the early Pliocene. To be sure, some contend that much earlier forms of fossil primates, such as *Propliopithecus*, found in the Fayum deposits of the Oligocene in northern Egypt, are directly ancestral to man. Others, basing their arguments not on fossil bones or teeth, but on blood serum protein evidence obtained from living species of primates, assert that the pongids and hominids split from each other a very few million years before the Pleistocene. The disagreement between those who stick to the fossil skeletal and dental evidence, and those who study protein molecules is far from settled at this time. *Ramapithecus*, found, as stated above, in both India and East Africa in late Miocene deposits, suggests that what is now mostly desert country between northeast Africa and the Indian subcontinent must have still been sufficiently moist to support such a primate forest-dweller. The canine teeth in *Ramapithecus* had become hardly more prominent than the neighboring incisors and premolars, whereas in most Old World monkeys and apes the canines are practically fanglike, especially in males. Moreover, the shape of the dental arch or arcade in *Ramapithecus* is more parabolic or

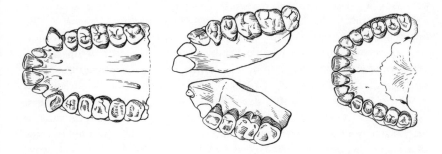

Figure 1. The dental arch or arcade and palate in (left to right) orangutan, *Ramapithecus*, and modern man. (After F. Clark Howell.)

V-shaped, as in man, rather than narrow or U-shaped as in apes. It has been pointed out that the narrow, U-shaped dental arch in modern apes may be a separate, late development, and that the parabolic arch may have been an old feature which man inherited. Just what selective forces would have brought about reduction of the large canine teeth, and this subtle modification in dental arch shape is not clear. Modern apes and monkeys employ their big canines not so much in eating as in fighting or in threat-displays connected with social dominance. For later hominid evolution, reduced canines would have been replaced, as far as the fighting or threatening function is concerned, by weapons used in the hands, but aside from one dubious bit of stone asserted to ·be a tool found in association with *Ramapithecus* in Africa, we have no evidence of tools or weapons at this early date in hominid development (Jolly 1972). Reduced male dominance might have rendered the big canines redundant, and there is a theory that body parts which are unused eventually diminish. This may sound like sheer Lamarckian evolutionary argument, but the point can also be made in Neo-Darwinian terms if it is assumed that any large body part (even a set of four big canine teeth) exacts a certain expenditure from the rest of the organism, so that if the big part no longer "pays its way" to offset this biological cost, it becomes a liability, so that individuals with mutations which diminish it have a survival advantage, however slight. It might stretch one's credulity to try to imagine what precise disadvantage extra large canine teeth might have, if they were no longer much used in fighting or even threatening, but they do after all lock up some calcium and other mineral elements which might otherwise be channeled into the growth of other teeth or the bones of the skeleton. Big fanglike teeth do get broken off either in fights or by accident and could lead to serious infection and ulceration of the tooth-socket and gum. This is a risk which the possessor of big canines must run, worth taking if they do provide some other advantages, but when they are no longer used, the liability is not offset. We might liken huge canine teeth to the oversized cowcatchers on nineteenth-century locomotives, which were highly useful in the days of open-range cattle and before the invention of barbed-wire fencing. Eventually they were made smaller, and the steel and labor effort the bigger ones required could now be utilized elsewhere. But this still does not tell us what was changing so as to produce these tooth and jaw effects in the earliest hominids though it is nice to suppose male aggressiveness was diminished. It is also possible to argue that man's ancestors never had excessively large canines, and that the massive canines seen in the pongids and Old World monkeys generally are secondary developments. In this view of the matter, it is unnecessary to explain the small hominid canines as the outcome of an evolutionary reduction process. Until we have a more complete set of fossil remains, this debate will continue without much chance of resolution.

A long controversy in anthropology has raged over what turned the ancestors of man into not only ground-dwelling primates (that happened also with the ancestors of the baboons, who remained quadrupeds), but upright or erect-walking ones (Napier 1970). The origin of hominid bipedalism could be the key problem of hominization, the critical change following which all the rest of man's distinctiveness became possible. Sixty years or so ago, it was believed by anthropologists that the impetus for hominization had been the enlargement of the brain, and that man's emergent intellectual prowess had triggered the other changes. The fossil record now shows that bipedal locomotion was achieved well before there was any spectacular enlargement of the brain, and that it was a new way of moving around rather than a need to handle more complex mental problems which set off human evolution in the strict sense.

Several hypotheses have been advanced to explain man's bipedalism. To begin with, it is clear that for this problem we need not explain the basic anatomical capacity to walk or run or stand bipedally, since apes and monkeys can already do these things. The point is that bipedal locomotion or standing is infrequent, not ordinary or habitual, with them. What has to be explained is how an *already present* but not often used behavioral capacity turned into a commonplace behavior. The bones and muscles of apes and monkeys permit bipedalism, but the effort is tiring after short distances, and if prolonged, perhaps even painful. The proportions of the various bones and muscles involved are simply unsuited for prolonged bipedality. Some monkeys and apes may run or walk upright when carrying bulky loads in their arms and hands. This can be seen in case certain large wild fruits, or even an armload of smaller food morsels, have to be transported (usually a short distance) away from others, in order to consume them undisturbed. Apes and monkeys do not store food, and are thus rarely faced with the need to move food of any kind very far from where they find it. The writer has suggested previously that a shift to bipedal carrying as a fairly regular thing may have been associated with a greater dependence on hunting or even scavenging the carcasses of game killed by other animals, when large chunks of meat (and bones) might have to be moved for longer distances in order to avoid predators or scavengers. It has been noted that apes and monkeys may rear up on their hind legs to get a better view of the landscape in areas of tall grass, although this momentary upright stance hardly seems more likely to make an animal into a regular bipedal walker or runner than the similar habit of many rodents, who stand or sit bipedally while surveying the scene. Further, apes and monkeys seem to dislike getting their hands on muddy, very cold or frosty ground, or very hot ground (as is the case with bare, rocky surfaces in the tropical sun). They also wade bipedally in shallow water. All this suggests that several factors combined may have made more frequent bipedalism useful, even if it meant sacrificing speed and balance (until sufficient structural changes in the hind limbs and pelvic region bones and musculature had taken place).

The anthropologist S. L. Washburn has suggested that there was a phase in which the ancestors of man walked quadrupedally on their knuckles, as do gorillas, chimpanzees, and orangutans, and that only later they rose from this halfway upright position to full bipedality. Other anthropologists are dubious about any "knuckle-walking" phase of this sort. It has also been suggested that the use of tools and weapons would have favored bipedal standing while wielding the implement, as well as bipedal walking while carrying implements from place to place. Aside from the fact that many tools are used while sitting on the ground or squatting, it must be admitted that effective use of a club or spear requires an upright stance, and that tool or weapon transport, like food transport, would have favored more regular adoption of bipedalism. It seems more likely to this writer that effective bipedalism, although not as well established as it now is in man, came well before any regular use of such weapons in hunting or human combat.

In any case, achievement of erect posture and bipedal gait entailed changes in the balance of the head on the neck, reducing the need for very heavy rear neck muscles, shifting the opening (foramen magnum) for the spinal cord into the braincase farther forward, so that it lies directly above the center of gravity of the body, and, in more complex fashion, altering the mechanical system of the jaw muscles. By taking the weight of almost half the body from the arms (in a quadrupedal position), and moving the entire support of the body weight to the hind limbs, major changes could be predicted in the size and arrangement of leg muscles. Among these results are the very prominent buttock and thigh muscles in man, compared to which the corresponding muscles in apes look positively scrawny. Eventually, also, changes could be expected in the foot bones and muscles, with a reduction in the flexibility of the great toe, and the development of the so-called arch which makes man's footprints so different from those of apes or monkeys. The pelvic bones have also been reshaped to handle the new position, so that the abdominal viscera in man come to sit in a shallow, basinlike pelvis, whereas the belly organs of the apes, in their normal walking or standing position, simply are suspended by the baglike abdominal wall and its muscles. The new locomotor position may also have made face-to-face or ventral copulation more comfortable (though it does occur at times among apes, especially the pygmy chimpanzee), instead of the usual dorsal or rump copulation posture common to most other quadrupeds.

Aside from the attraction of game animals feeding in large flocks or herds on the grassy plains to a primate-turned-hunter, it has been suggested that the grass itself was being used for food, especially the small grass seeds which are the wild counterparts of our domesticated cereals. Baboons eat young and tender grass shoots, but certainly do not subsist regularly on mature, ripened grasses. Even if the emerging hominids were eating the seeds and tender young wild grasses, to an extent far greater than any of the apes or monkeys, they obviously did not go on from there to develop the digestive tracts and specialized molar teeth which would have enabled them to subsist on the grass in the manner of the herbivores. The leaves of mature grasses are exceedingly hard, and actually contain abrasive amounts of silicon, which wears the molars down quickly, unless these teeth consist of pleatlike folds of enamel so that when the surface is worn down, there are still fresh, resistant enamel ridges. The hominids managed to partake of the bounty of the great grasslands in other, less direct ways. The first way is to eat the grass-eaters. We still do this. Man's long reliance on hunting—mostly of grazing or browsing ungulates—avoids the necessity of undergoing the drastic modifications permitting direct feeding on grass or woody browse. A later way was to harvest the seeds or grains, and to make them digestible for human stomachs and intestines by the pounding or grinding them into flour or meal, and then cooking the result in a variety of ways, from pancakes or porridge to bread, or making the mash into beer. Although early man probably learned how to harvest wild grains, it was not until agriculture was invented that grass seeds became a dietary staple. Still another way of turning grass into food is to milk cows and goats and other ungulates, and then either to drink the milk or to process it still further into butter or cheese. Grass seeds can also be used to feed chickens, producing both chicken-flesh and eggs.

In very broad perspective, it is apparent that man's enormous numbers in the modern world depend on getting food out of various (now domesticated) grasses or the scantier grasses of natural pasturage. About 70 per cent of the caloric intake of the modern Chinese comes from grass seeds, in the form of rice, wheat, and millet. Although the modern American direct cereal intake measured in calories is much less (around 20 per cent), by the time we add the foods indirectly obtained from grasses, that is, most of our meat, not only beef and lamb, but even pork, since it is mainly corn-fed, our milk, butter, cheese, eggs, corn oil, our own diet is not much less grass-dependent than that of the Chinese. We do have other sources of foods such as starchy tubers, fruit, legumes, fish, etc., but most of humanity would be starving if man had not learned how to extract the bulk of his diet from grass. Even sugar, which makes up a significant part of our modern caloric intake, is mostly obtained from a cultivated grass (sugarcane).

If bipedalism proved to be the trigger for the anatomical changes which stamp the hominids as a new and different kind of primate, the predisposing cause of the shift to the new locomotor pattern was almost certainly connected with the food supply in some way or another, so that the new way of moving about and carrying things enabled our ancestors to start exploiting food resources theretofore largely neglected by the primates.

Among the factors which appear to have played a large part in the unusual evolutionary history of mankind are some which are in fact common to primates in general, or at least widespread among the higher primates of the Old World monkeys and apes. The young of primates (like those of several other kinds of mammals) are born in a fairly helpless condition, unable to walk, though not, like puppies or kittens, blind. This is what is called *altricial*—having helpless infants. The opposite condition is where the newborn are soon capable of moving around, feeding by themselves, etc. Young chicks are precocial, whereas the nestlings of songbirds are altricial and have to be fed by their parents. All mammals are nursed by their mothers, to be sure, but in addition, primate mothers must carry their feeble offspring about for a while, since they can only crawl ineffectually on their own. In addition to being altricial, the higher primates have fairly long lives: macaque monkeys live twenty-five to thirty years, and the great apes may reach the age of forty or more. It has been observed that whether animals are altricial or precocial, a long lifetime affords much greater opportunities for learning, whether individual or social. Length of life is not closely correlated with duration of infancy and childhood, or the age of sexual maturity. Modern man has a considerably longer period of immaturity than his average life span might suggest. In any case, prolonged immaturity, whatever disadvantages it may have, also seems to allow for more learning. Although a longer childhood might also allow for the brain to continue growing (until the skull sutures close or fuse), human brain growth levels off quite soon, and the brains of four- or five-year-old humans are not all that much smaller than those of adults.

We have already noted that the primates generally (above the prosimians) have excellent eyesight, compared to that of other mammals, resembling that of the birds (Young, J. Z. 1971). Simply living in trees does not confer superior eyesight, since there are many other mammals living in the trees who are not outstanding visually (koalas, tree sloths, fruit rats, squirrels, etc.). The primate visual abilities are evidently connected with extreme skill in coordinating the hands (and feet) in both climbing and leaping about among the branches, finding food mainly by sight and carefully peeling or handling it with visual inspection, and careful grooming of the bodies of others as well as of the self. Primates appear to recognize each other (again, excluding the prosimians) mainly by sight, rather than smell. It is somehow widely believed that only man possesses an "opposable" thumb, but observation of apes and monkeys shows that they can hold or grasp things between thumb and fingers fairly adeptly, even though man is by far the most skilled finger-manipulator among the primates. True, the gibbons have almost vestigial thumbs, and the spider monkey of South America has no thumbs at all.

Many writers on human evolution have emphasized the juvenile or infantile character of man, and have proposed that human evolution has been a kind of delayed maturation—sometimes called *neoteny* or even *fetalization*. Such features as sparse body hair, small jaws, small canine teeth, a rounded or even bulging skull, and in the area of behavior, a seemingly greater tendency to remain playful and curious, have been cited as supports for such a view. On the other hand, man has numerous features which are not at all juvenile, such as long legs, prominent buttocks, prominent nose, regular bipedal locomotion, and a noninfantile vocal tract. Further, some of the features described as "neotenous" are features found only in recent mankind—during the last 40,000 years or so, or for only a fraction of 1 per cent of the time that the hominids have been around.

Having only one baby at a time was mentioned as a general primate characteristic—and certainly the major reason why primate mothers have only two nipples rather than four or six or more. With only one very young infant at a time, the primate mother can devote most of her attention to its care; with the advent of man, and a still further prolongation of infant and childhood dependency, this meant an even longer period for the child to learn behavior from its mother, or to engage in play with other young individuals of the same local social group (Eimerl and DeVore 1965). Most primates live in fairly large social groups (as do a good many other mammals, to be sure).

However, their relatively slow passage from infancy through childhood, and the prolongation of mother-child concerns beyond the year or so characteristic of other large mammals (except such huge and exceptional mammals as elephants and whales), creates for the primate social group a set of unusually favorable conditions for social learning (Dolhinow 1972). Young chimpanzees in the wild have ample opportunity to observe the behaviors of elders, even though the elders seem to make no effort to "teach" the young (Goodall, 1971). The use of a simple twig tool to extract termites from termite hills by chimpanzees has apparently been transmitted in this manner. We do not know when the first hominids started to teach their children, rather than simply let them continue to learn by example, but it is possible that a fair amount of crude tool-making and tool-using, and food preparation such as butchering, might have been socially transmitted without deliberate instruction of the young. The most important single behavior learned by human children—the use of language—occurs in most cases without formal instruction.

Early Man
and Hunting

With man's having "come to the ground" and adopted a bipedal gait, and perhaps having come to depend much more on hunting than the chimpanzees of today, where males occasionally kill some small mammal and eat it on the spot, we must consider some new problems. Sherwood L. Washburn and C. S. Lancaster have suggested some of the relationships of early hominid hunting to the rest of early hominid life:

The success of the human hunting and gathering way of life lay in its adaptability. It permitted a single species to occupy most of the earth with a minimum of biological adaptation to local conditions. The occupation of Australia and the New World was probably late, but even so there is no evidence that any other primate species occupied more than a fraction of the area of Homo erectus. *Obviously this adaptability makes any detailed reconstruction impossible, and we are not looking for stages in the traditional evolutionary sense. However, using both the knowledge of the contemporary primates and the archeological record, certain important general conditions of our evolution may be reconstructed. For example, the extent of the distribution of the species noted above is remarkable and gives the strongest sort of indirect evidence for the adaptability of the way of life, even half a million years ago. Likewise, all evidence suggests that the local group was small. Twenty to fifty individuals is suggested by Goldschmidt (1959:187). Such a group size is common in nonhuman primates, and so we can say with some assurance that the number of adult males who might cooperate in hunting or war was very limited, and this sets limits to the kinds of social organizations that were possible. Probably one of the great adaptive advantages of language was that it permits the planning of cooperation between local groups, temporary division of groups, and the transmission of information over a much wider area than that occupied by any one group. . . .*

When males hunt and females gather the results are shared and given to the young, and the habitual sharing between a male, a female, and their offspring becomes the basis for the human family. According to this view the human family is the result of the reciprocity of hunting, the addition of a male to the mother-plus-young social group of the monkeys and apes.[7]

David Pilbeam adds some additional points to this picture of man the hunter:

The adoption of hunting would have produced for the first time in primate history a division of labor between the sexes. The females retained their roles as gatherers of vegetable food (also, of course, they were responsible for care of young), while the males were hunting. For hominids, hunting, as for the small carnivores such as wolves and hunting dogs, requires cooperation among a number of individuals. Not only would the hominid troop be split up when most of the adult males went on hunting expeditions, but for the first time cooperative behavior among males (and also of course among females) would become essential. As a further corollary, the shift to cooperative hunting would of necessity require food sharing between males and females and probably also among subgroups within the troop. Both cooperative behavior and food sharing are atypical behavior in nonhuman primates, occurring only occasionally in the chimpanzee.

With the males away on periodic hunting expeditions, hominid troops would have ceased moving their sleeping area each night as the vegetarian foraging primates do but instead would have set up temporary campsites for periods of several days. Thus there was a base camp for the males to return to with game and a place for butchering, sharing, and eating the meat over a period of days.

The importance of hunting has perhaps been overemphasized in discussions of hominid evolution; undoubtedly, many of the stone implements used by hominids were meat-preparing tools, yet the technology involved in plant collection and particularly in its preparation was perhaps equally important. Much of the plant food available in open country requires quite complex preparation—crushing, soaking, and so forth—and this aspect of the hominid's developing technological skills should not be overlooked.

The growth of cooperative behavior among members of the troop would have produced a change in dominance interactions, particularly among males. This required not only suppression or control of aggressive behavior but also the development of new neural structures for positively mediating

---

[7]S. L. Washburn and C. S. Lancaster. 1968. "The evolution of hunting." In Richard B. Lee and Irven Devore, eds., *Man the hunter.* Chicago: Aldine Publishing Company. Also reprinted in Phyllis Dolhinow and Vincent M. Sarich, eds., 1971. *Background for man.* Boston: Little, Brown & Company, pp. 387-405.

cooperative activities. The reduction of male aggression would have facilitated their integration into the troop in a positive way, perhaps through the development of permanent pair bonds between male and female.[8]

The tropical savannas and several other such environments not only teemed with hoofed game, but were the home of big and medium-sized predators—members of the cat family such as lions and leopards, and wild dogs, hyenas (not, as widely supposed, only scavengers), presenting some risk to bipedal primates much slower on their feet than most of the open country hoofed game. Most people have quite erroneous ideas about predation. Under most natural conditions, predators do not cause the extinction of species on which they habitually prey. Instead they provide one kind of violent Malthusian population control, eliminating chiefly the sick and surplus young, or the very old. Most predators tend to specialize on prey where the payoff is most predictable and avoid more unpredictable creatures, which early hominids must have been. The late L. S. B. Leakey, with wide experience in East African game areas, contended that the flesh of hominids is unpalatable and that it has a bad smell and would not be eaten except by very desperate, hungry predators. The fact that people are still killed by predators reflects a number of special conditions which would not have been very frequent in the prehistoric past, including victims (most often women or children) using forest trails, or males hunting alone who come too close to a big predator but fail to get in a killing shot. Recent kills of human beings by grizzly bears in North America have been of individuals in very small groups who made the mistake of sleeping out in isolated spots. It seems very unlikely that the hominids ran enormous risks in venturing out into the open country. As long as they camped in fair-sized groups or, as hunting parties of males, carried even sticks and stones, they were probably generally safe. When man began to keep fires, safety increased still more. It is not quite clear why predators (or other animals other than some of our domesticated ones) seem to fear or avoid fires; for animals guided in their behavior by scent, a smoky camp would probably be confusing.

Prior to the invention of specialized weapons, the early hominids may have killed game of manageable size by neck-breaking. Heavy branches or large rocks could be used to bash skulls. Raymond Dart, who described the first South African "man-ape," believes that leg bones, especially the femur, which has a knoblike upper end, would have served as effective clubs for killing game. If this seems an unpromising catalogue of weapons, all of which

---

[8]David Pilbeam. 1972. *The ascent of man: an introduction to human evolution.* New York: The Macmillan Series in Physical Anthropology, pp. 152-3.

require getting within a few feet of the prey, it should be remembered that the victims may have been, more often than not, already injured stragglers from the herd, infant ungulates lying quietly in concealing vegetation, or animals too worn out from dogged pursuit to run further. Projectile weapons, enabling animals to be killed at a distance, came much later, to say nothing of traps, snares, deadfalls, pitfalls, etc. Another hunting method, widely used in the past by North American Indians and in prehistoric Europe, was the drive of a herd over a cliff or bluff, which might yield enough badly injured animals to make killing easy. While the Indians used bows and arrows, or spears, and organized such "jump" kills to the point of constructing enclosures at the foot of cliffs over which herds were stampeded, some return might be expected from a much less well-prepared drive.

It cannot be assumed that the increased hunting (for which we have some archeological evidence) was the only food source (Lee and DeVore, 1968). Most reconstructions of the way of life of early hominids assume that while the adult males did the hunting—at least of the large game—the females and young collected wild plant foods. This latter activity may have differed from that of the plant-collecting activities of apes and monkeys, where all take part—males, females, and young—in that nonhuman primates do not share food with one another. Very limited instances of food-sharing have been seen in monkeys and in wild chimpanzees who simply gather food and consume it then and there. Nothing is carried for any distance, and nothing is stored "for a rainy day." If the model we have of small hunting parties of hominid males and plant collecting carried on by females and children is to be workable, food-sharing must be part of the picture. This means that the hunter males, instead of simply eating their kill where they made it, must have hauled parts of the carcass "home" to camp, where at least some of the meat would have been shared with females and young. Likewise, unless some of the collected plant foods—edible roots, shoots, leaves, berries, fruits, nuts, or seeds—were carried home and shared with the males, there might have been considerable periods of food shortage for the males—unless they too shifted to plant collecting when game was scarce. To account for such drastic modification in what seems to us as the highly selfish feeding habits of nonhuman primates, except of course for the nursing of infants by their mothers found in all mammals, is definitely a tough problem. Special attention to the young was not entirely a matter of female behavior; males, too, among the primates, are concerned with infants, as Alison Jolly notes:

Males' attitude to babies ranges from indifference in some species such as langurs to performing the bulk of the care in marmosets. Males of most species will allow infants to take liberties and will protect them from outside threat. In baboons, macaques, and sifakas, infants form a focus of interest for males and females. Male barbary apes and hamadryas baboons carry infants to gain status for "social buffering."

Females and female juveniles groom and carry infants not their own, so-called aunt behavior. This seems common in all social species and may be an important way for female juveniles to learn appropriate maternal care.[9]

It may be no coincidence, however, that food-sharing of *meat* occurs among some carnivores and that the Cape Hunting Dog (*Lycaon pictis*), a wild African pack-hunter, brings back meat from a kill and regurgitates it for whatever animal has remained with the pups. To be sure, there is no adequate theory of why the Cape Hunting Dog has acquired this unusual behavior. A pack-hunting, carnivorous way of life in an already highly sociable species probably, in some manner we do not fully understand, favors food-sharing behavior.

We have also spoken of a home "camp." Most primates move about, sleeping from night to night in different trees, although a few baboons frequent rocky cliffs which provide equally safe sleeping places. No primate other than man maintains a camping place for several days or more at a time, to which the members return from various food-foraging expeditions. Some camping places may have been in rock-shelters or at the mouths of caves, but such spots are quite rare. Most would have been out in the open, although perhaps near suitable trees in the unlikely event of a predator attack, or more simply, for shade in the heat of the day. This brings up the point that there may be more than an accidental relationship between open-country hunting treks and man's generally scanty body hair. Apes and monkeys are rather hairy or even furry, but they are active mostly in forest shade, and spend little time in bright sun. Baboons and the patas monkey are occupants of sunny, open environments, and have fairly short hair (except for a kind of mane in adult male baboons). But even baboons and patas do not pursue game in the heat and sun. It has been suggested that lengthy game pursuit would have produced overheating and risk of heatstroke, unless offset by the cooling effect of heavy sweating preferably without a layer of insulating body hair. If the females did not hunt, one might wonder how the females would have lost their furry coats, except that selectively favorable mutations will appear in both sexes unless the responsible gene happens to be on only one of the pair of sex (X and Y) chromosomes. Unfortunately the fossil record sheds no light on the problem of when mankind began to lose his body hair, which

[9]Alison Jolly. 1972. *The evolution of primate behaviors*. New York: The Macmillan Company, pp. 266-7.

led to the condition referred to by Desmond Morris as *The Naked Ape* (1967).

We have spoken of the new, meatier diet, simply as an opportunity to make use of an abundant but harder-to-get food. But meat is high-quality, high-energy food, heavy in protein and also animal fat, in contrast to the rather low-grade fodderlike diet of most primates except when they can eat ripe fruit or rich nuts. Although there are vegetarians in the modern world who believe that most of man's ills, even his aggressive behavior, can be blamed on eating meat, it is more likely that the addition of regular amounts of meat to the hominid diet had several real advantages, including greater resistance to certain diseases, diminished infant mortality, and greater longevity. The longest-lived modern peoples are those whose diet is very high in animal protein, and the shortest lived tend to be the most completely vegetarian. Protein deficiency in extreme cases produces an often fatal condition in young children called *kwashiorkor*. Less extreme protein deficiencies, if prolonged, may cause mental retardation. It may seem simplistic, but it is possible that a switch to a meatier diet enhanced the mental capacities of the hominids, although other factors were probably more important.

If significant amounts of animal carcasses were brought back to a home camp or base, it would follow that the camp area would be littered with bones, animal horns, teeth, and pieces of hide and tendon. Such material could have served as the basis for the "osteodontokeratic" (i.e., bone, tooth, and horn) tool-kit attributed to Australopithecine hominids by Raymond Dart, although almost equally usable tool material would exist in most localities in the form of broken branches, sticks, or pebbles and cobblestones. The living floors of early hominid camps found in the lower levels of Olduvai Gorge, Tanzania, contain remains of some small creatures—frogs, snakes—and not just the bones of the larger ungulates. If we had evidence of shoreline living sites for hominids at the same period, they would contain remains of shellfish as well.

Among the new pressures on open-country, game-hunting hominids there would have been some new diseases, especially those transmitted by insect bites from one warm-blooded animal to another. Sleeping-sickness (encephalitis lethargica) may have been one of these diseases, spread by the bite of tsetse fly. On the other hand, the early hominids, not crowded into unsanitary permanent villages, but living in mobile bands or troops, would have suffered less from fleas, lice, leprosy, yaws, cholera, tuberculosis, plague, etc., than many modern peasants.

If the pattern we have been describing actually did prevail for a long time among the early hominids, there would have been some additional side-effects (cf. Spuhler 1959). Apes and monkeys characteristically occupy small ranges or territories, and in their wanderings rarely get very far from each other; baboons, which move about considerably more than most other primates, tend to move in a troop, with individuals seldom out of earshot, and usually within visual contact. As game-hunters, however, hunter-hominid males must have often wandered many miles from the "home" group, perhaps for two or three days at a time. Success in this way of life would depend on the ability to find the way back to camp—in other words to develop enough trail sense to avoid getting lost. One of the obvious pressures would have been on memory for direction and places, almost wholly based on vision (whereas with most other mammals, direction-finding is more a matter of smell). Psychologists have noted that in modern subjects, boys show better spatial and directional skills than girls (just as girls are consistently superior to boys in verbal skills). This may be due merely to cultural influences, but it could reflect an old, innate sex-difference, which would be consistent with the presumed greater wandering propensities of hominid males for perhaps two or three million years.

Although meat eating seems to have increased during the past millions of years of hominid existence, there is some dispute about the intensity of hunting. It has been argued that, just as among recent primitive hunter-gathering peoples, plant foods would have really continued to be the dietary mainstay (Lee and DeVore 1968). The proponents of a long period of concentrated hunting argue that many of the vegetable foods now gathered are poisonous, or at least virtually indigestible, without cooking or other forms of processing presumably not invented until comparatively recent times. Cooking could not have appeared before man used fire, and probably not much more than half a million plus years ago (on the basis of finds in Hungary and southern France); further, the kinds of cooking needed to render some of the wild plant foods (or their domesticated variants) edible usually involve use of boiling water, not easy to achieve without pottery. True, one can boil water by digging a hole, lining it with an animal skin, and heating the water with red-hot stones pulled out of the fire (this is known as "stone-boiling"). We have no evidence as to the antiquity of this method of circumventing the absence of cooking pots.

The
Australopithecines

Despite Leakey and others who see in *Ramapithecus*, a fossil form dating back perhaps as far as 20 million years, but more likely about 12 million years, the precursor of man in a direct sense, it is not until we come to the fossil creatures called *australopithecines* that the picture looks fairly clear. The first *Australopithecus*, named because it was found in South Africa (it has nothing whatever to do with Australia) in 1924, was recognized by Dr. Raymond A. Dart, then teaching anatomy at Witwatersrand University in Johannesburg, as a new manlike ape or apelike man (Dart, 1959). Dart's own account of his reactions as he unpacked the first specimen which he named *Australopithecus* is worth quoting:

I wrenched the lid off the first box and my reaction was one of extreme disappointment. In the rocks I could make out traces of fossilized eggshells and turtle shells and a few fragmentary pieces of isolated bone, none of which looked to be of much interest.

Impatiently I wrestled with the lid of the second box, still hopeful but half-expecting it to be a replica of its mate. At most I anticipated baboon skulls, little guessing that from this crate was to emerge a face that would look out on the world after an age-long sleep of nearly a million years.

As soon as I removed the lid a thrill of excitement shot through me. On the very top of the rock heap was what was undoubtedly an endocranial cast or mold of the interior of the skull. Had it been only the fossilized brain cast of any species of ape it would have ranked as a great discovery, for such a thing had never before been reported. But I knew at a glance that what lay in my hands was no ordinary anthropoidal brain. Here in lime-consolidated sand was the replica of a brain three times as large as that of a baboon and considerably bigger than that of any adult chimpanzee. The startling image of the convolutions and furrows of the brain and the blood vessels of the skull was plainly visible.

It was not big enough for primitive man, but even for an ape it was a big

*bulging brain and, most important, the forebrain was so big and had grown so
far backward that it completely covered the hindbrain.*

*Was there, anywhere among this pile of rocks, a face to fit the brain? I
ransacked feverishly through the boxes. My search was rewarded, for I found
a large stone with a depression into which the cast fitted perfectly. There was
faintly visible in the stone the outline of a broken part of the skull and even
the back of the lower jaw and a tooth socket which showed that the face
might still be somewhere there in the block.* [10]

There are no wild apes in South Africa today, so it would have been an
interesting find even had it been simply an ape. However, as Dart began to
study it, he found features indicative of a more manlike and less apelike
creature. When the jaws were separated and the teeth could be seen, they
were seen to be set in a parabolic dental arch, and although the specimen was
a child, it was assumed that when an adult specimen was found, it would
prove to be quite different from any living ape. Dart's prediction came true.

In the 1930s, a paleontologist named Robert Broom began finding more,
and fully adult australopithecines in South Africa, not far from
Johannesburg. The finds occurred in cavelike grottoes subsequently refilled
with limestone deposits, yielding also large quantities of broken mammal
bones. Eventually, very crude stone tools were found in a part of one of these
sites (Sterkfontein). Similar finds were made further afield, in the northern
Transvaal (Makapansgat), and it became apparent that at least two forms were
represented—a rugged, robust form which was so named (*Paranthropus
robustus*), and a more gracile, but larger-toothed form, which seemed to be
the adult of the type first described in 1924 by Dart (*Australopithecus*).
Fragmentary remains of a third form seem to have belonged to a later human
group, *Homo erectus* (see below). Dart (1959) concluded that the broken
bones had, in some instances, been used as tools or weapons, and named this
the "osteodontokeratic" tool-kit. Others have not been convinced that these
had been in fact early tools or weapons.

In East Africa, the late L. S. B. Leakey had been working for many years
as an archaeologist and paleontologist, but it was not until 1957 that his wife,
Mary Leakey, discovered the first of a series of East African examples of the
australopithecine group, at first named "Zinjanthropus boisei." In fact,
Zinjanthropus, with massive teeth (Leakey also called him "nut-cracker
man") except for his small incisors and canines, must have had a massive jaw;
his huge jaw muscles were inserted along the top of his head in a back to
front crest of thin bone. Closer study indicates that "Zinjanthropus" is not

[10]Raymond A. Dart (with Dennis Craig). 1959. *Adventures with the missing link.*
New York: Harper & Brothers, Publishers, pp. 4-7.

very different from the South African robust *Paranthropus*. As usual, there has been a confusing tendency for the discoverers of fossil human remains to invent new names for each find. Later, Leakey was more careful about this.

The "Zinjanthropus" find had been made in the deep canyon known as Olduvai Gorge, in Tanzania, west of the Great Rift Valley, on the edge of the great game reserve of the Serengeti Plain. Further work in Olduvai Gorge soon turned up other finds from the same geological time interval, between about two million and one million years ago. Leakey discovered a much less robust, but apparently larger-brained form he named *Homo habilis* (skillful man), which resembles the original *Australopithecus* more than the *Paranthropus* "Zinjanthropus" form. A few anthropologists have insisted that these different forms represent the male and female sexes of the same species. This notion, propounded by C. Loring Brace, does not have much support from other scholars. One of the things which has characterized most of the Old World Cercopithecoid monkeys, and, aside from the gibbons, the anthropoid apes, has been marked sexual dimorphism, that is, a sharp difference in size and other features between males and females. Adult male baboons are almost twice the size of adult females. Male gorillas are considerably bigger than females. Sexual dimorphism is less extreme in chimpanzees, and is scaled down even further in man.

Annoying issues are raised if two different hominids coexisted in southern and eastern Africa in the early Pleistocene, or late Pliocene (Washburn 1963; Pilbeam 1972). There is a well-known biological principle of "competitive exclusion," which argues that any two (or more) closely related species with very similar subsistence and other requirements cannot long occupy the same territory. For coexistence to be feasible, there must be some differences in subsistence behavior. However, when one contemplates the still remaining huge herds of several kinds of hoofed mammals, all grass-grazers, which occupy areas of East Africa today, it is hard to see what subtle behavioral differences sustain the principle of competitive exclusion; these animals have eaten the same grass, frequented the same waterholes, for a very long time.

After the discovery of "Zinjanthropus" at Olduvai, a large lower jaw which belonged to a very similar individual was found at Peninj, to the north; this jaw has been used in making the reconstruction of the Olduvai form, which when found lacked a lower jaw. Still more recently, numerous fragmentary finds of australopithecines have turned up west, north, and east of Lake Rudolf, in northern Kenya. The potassium-argon dating of some of these finds indicates that they existed as far back as 5 million years ago. How far back the australopithecines go is still an open question, but they evidently date to the middle Pliocene. This still leaves a major gap of several

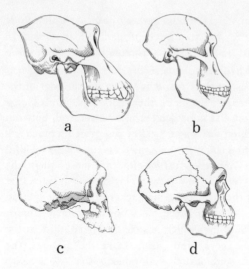

Figure 2. Lateral views of skulls: a. female gorilla; b. "plesianthropus," an *Australopithecus* found at Sterkfontein, S. Africa; c. KNM-ER 1470, a more advanced hominid found east of Lake Rudolf, Kenya, in 1972; d. *Homo erectus* (reconstructed by F. Weidenreich), near Peking, N. China.

million years if their predecessor was *Ramapithecus* or a similar form. An investigator who has worked with blood biochemistry or serological tests of biological affinity, Sarich (Dolhinow and Sarich 1971: 60-81), believes that the separation of the australopithecine from the pongid line occurred not much longer ago than five or six million years, but those who put more faith in the evidence of bones and teeth remain convinced that the separation is twice that old.

In the lowest beds at Olduvai Gorge, Leakey found some exceedingly crude stone specimens which seem to have been rather casually chipped, consisting of choppers and chopping tools, which he designated as the "Oldowan" tool tradition (Bishop and Clark 1967; Howell and Bourliere 1963). Similar very crude stone tools, or possible tools, had previously been found and described as "pebble tools," although they were not all necessarily made from stream pebbles or beach pebbles. This tool-kit of stone was almost certainly not the whole of australopithecine technology. Wooden tools such as very rough clubs and sticks were probably used. It is also conceivable that some of the simpler uses of cords or thongs had already been devised for carrying loads of meat, or even simplifying the carrying of tools by means of cord around the waist, leaving both hands free. Vegetable foods gathered by the females may already have been hauled home in rolls of tree-bark, or containers of large leaves. Leakey found a vaguely crescent-shaped pile of stones near the base of Olduvai Gorge which could have been part of a very rough shelter. It seems clear, however, that the australopithecines made no fires.

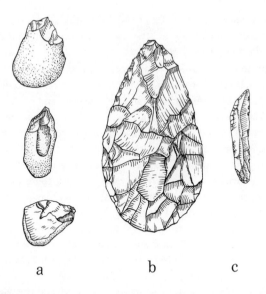

a   b   c

Figure 3. Successive levels of stone tool-making skills: a. Oldowan tools, E. Africa; b. Acheulean handaxe; c. Upper Paleolithic blade implement.

It would be important to know whether or not these creatures already had achieved language (Pfeiffer 1972). They could communicate, since all animals communicate in some fashion, but the presence of true language means that propositional statements about the environment are possible. Apes and monkeys communicate by means of vocal calls, facial expressions, and some body movements or gestures, mostly about their feelings (Goodall 1971). A few monkeys can convey information about whether predators are approaching on the ground or from the air. It is possible that apes can convey some simple messages: for example, that they have encountered a snake. Recent experiments with chimpanzees in an open-air enclosure yield surprising amounts of cooperative behavior which implies some kind of social communication. It was formerly assumed that the smaller brains of apes meant that they could not handle "symbolic language," but recent studies with the chimpanzees Washoe, by R. A. and B. Gardner, and Sarah, by David Premack, have shown that when taught *nonvocal* language codes, they can handle at least a modest vocabulary and grammatical system. It therefore seems clear that the australopithecines, with brains about the same size as those of modern chimpanzees, could have handled a nonvocal language, provided it had developed, and they had an opportunity to learn it by observation and imitation. You will note, however, that it has not been possible, so far at least, to get apes to use vocal language—speech.

Experimental reconstructions of the vocal tracts of australopithecines, based on fossil remains, and on some later fossil hominids of the *Homo erectus* grade, by Lieberman, Crelin, and Klatt, indicate that vocal language—speech resembling present-day spoken languages with various precisely articulated vowels and consonants—would not have been physically possible. Speech of course is both a matter of mouth parts and larynx, and of the brain. In view of these considerations, the writer has tentatively concluded that at this epoch in hominid evolution, propositional language was mainly a matter of nonvocal manual and arm gestures. Such vocal cries or calls as were uttered served mainly for emotional expression, or emphasis of messages being gesturally communicated. If this hypothesis is valid, there must have been a transformation from mainly gestural to mainly vocal language. This may not have occurred until the Middle Paleolithic.

There has been much interesting speculation about the origin of the human family system, which in many ways is unlike anything in the rest of the Primate Order (Morris 1967; Montagu 1965, 1968; Washburn, 1961). Apes and monkeys are generally sexually promiscuous, and although maternity is recognized by both mothers and offspring, apparently for many years or lifelong, in the sense of a special social relationship, monkeys and apes have no comparable attachment to a particular male as "father." Recognized paternity is a social phenomenon of man. The gibbon is the only "monogamous" primate; more commonly a male may mate with several females, and among some species of baboons, this may involve much jealousy and the maintenance of what have been called harems (Kummer 1971). In human societies, despite the many variations in the marital state, paternity is recognized, and in the majority of families, a male and one or more females are "bonded" in a close association which includes their offspring, usually for several years at a time. Rough analogies to pair-bonding occur in some other mammals, and in many bird species, but the relationship does not seem to arise in the ways that it does among human beings. The utility of such pair-bonding is that the male may serve as an additional protector for the offspring and, in some cases, an additional food-provider; the male may defend the surrounding territory from intruding males of the same species, and so on. Under some circumstances, such arrangements apparently enhance the survival of the offspring and hence of the species as a whole. In hominids, one key to the pair-bond is the practice of food-sharing, found in all human societies today, and probably very ancient. Another factor is the suppresssion of incest—the mating of close relatives such as mother-son, father-daughter, or brother-sister. All human societies forbid such matings, at least as a general rule. It was formerly asserted with confidence that incest-avoidance was absolutely peculiar to man, and absent among all other animals. However, recent close observation of some monkeys reveals that mother-son sexual relations are rare or absent, although no other forms of "incest-avoidance" seem to occur. The disinclination for sex between mother and son seems to arise from the established dependency which, in primates, is remembered for years or to the end of life, and constantly renewed by such activity as more frequent grooming of sons (and daughters) by mothers, and from the fact that male sexual behavior involves an assumption of social dominance incompatible with the dependency bond established and maintained between mother and son. Since fatherhood is unknown and unrecognized, no similar incest avoidance appears to prevent father-daughter sex; the problem of why sibling mating should be suppressed among humans is not solved by these findings and speculations. It has been observed that a few simple facts having to do with age and sexual maturation would make some of these "incestuous" relationships rare. In the case of brothers and sisters, one or the other would normally reach sexual maturity well before the other, and given a high rate of infant mortality, the age gaps between siblings would be more likely to be on the order of two or three years than closer. When one sibling attained sexual maturity, he or she would find a mate of suitable age from some other family, thus reducing the chances of a brother-sister pairing. Likewise, in addition to the mother-son relationship already discussed, sons would not reach sexual maturity until their mothers were already comparatively old. We have no real warrant for assuming that younger "nubile" females would, for earlier hominids, appear to be more sexually attractive than somewhat haggard, aging females (their own mothers), but it remains a plausible assumption, because it may be easier for a young male to establish sexual dominance over a just-maturing female than over an older, experienced female who has already had a previous mate or mates.

It has been suggested in connection with early hominid family life, that a pair-bonding system would allow male hunting parties greater peace of mind on their journeys away from base-camp than a system in which promiscuity prevailed, but this implies that early hominid males were as easily aroused by sexual competition from other males as are adult male baboons. Chimpanzee males, it turns out, are seemingly not bothered at all by sexual advances made to females with whom they have been intimate. Faith in the pair-bonding system, or fear of "adultery" or "cuckoldry"—obsessions in many recent societies—may in fact have played no important role at all among the early hominids. Concern about such matters may not have appeared until long after language had developed to the point of supplying emotionally charged abstract symbols which could arouse as much violence as witnessing a real social event.

The demography of the australopithecines has been the subject of some speculation. There have been suggestions about the size and composition (in terms of sex and age groupings) of local bands or troops of these creatures (Washburn 1961). Another question, equally impossible to answer, is how populous the entire species might have been at a given time. If we judge from the numbers of modern gorillas, orangutans, or chimpanzees, we might conclude that total numbers were very small. However, we must recall that the geographic range of these creatures seems to have been much wider than that of these modern apes, and that suitable environments may have covered larger areas of the Old World in the late Pliocene and most of the Pleistocene than they do nowadays. A half million individuals might have existed at the same time, probably divisible into several local varieties or "races." Spread all the way from South Africa to India or Indonesia-Southeast Asia, this would not mean that they were especially dense. Existence of australopithecines in Indonesia rests on a problematical fragment of a lower jaw found in Java called *Meganthropus*.

Homo Erectus

Among 1972 finds from east of Lake Rudolf, Kenya, Richard E. F. Leakey, a son of L. S. B. Leakey, reported a specimen designated KNM-ER 1470, consisting of a cranium and facial fragments, with an estimated cranial capacity of over 800 cc, dated at about 2.9 million years ago. On the basis of brain size and some other features, this specimen appears to be close to the grade of man known as *Homo erectus*, heretofore considered to date no earlier than about 1 million years ago (Leakey, R. E. F. 1973; Hooton 1946; Brace et al. 1971; Day 1965; Leakey, L. S. B. and Goodall 1969; Brace and Montagu 1965). The first find of this group was made in 1894, by Eugene Dubois of the Netherlands, in eastern Java, and for a long time the find was known as "Pithecanthropus erectus" (the erect ape man), or, more simply, Java Man (Osborn 1915[1922]). The "erectus" referred to the long, straight thighbone found with the skull cap, some teeth, and nothing else in a gravel bed along the Trinil River. Years passed with no further discoveries and Dubois himself ended up rejecting the hominid character of his find, insisting that it had been a kind of giant gibbon. However, in 1907 a massive lower jaw, now believed to belong to a member of the same group, was found in a Rhineland sandpit near Heidelberg. In the Late 1920s, a number of skulls and some other bones, along with remains of both very crude stone tools we would now describe as more or less Oldowan, and also with traces of fire, were found in a limestone-filled cave deposit southwest of Peking, in North China. This was dubbed "Sinanthropus pekinensis," implying still another genus of fossil hominid. This is now considered to be somewhat more advanced form of *Homo erectus*. In the thirties, further remains of Java man came to light in Java, and still more remains have been discovered since World War II by Indonesian scientists. Further remains of *Homo erectus* have since been found in northwest Africa, in Morocco and Algeria, and at Olduvai Gorge, higher up than the "Zinjanthropus" and "Homo habilis" material. Plotted on a map of the Old World, this indicates that many had by now (roughly from some time after one million years ago to about 350,000 years ago) achieved the occupation of a very large area, within which were many different kinds of climate and vegetation. Small skull fragments and tools of the same "Oldowan" type, with fire, were found still more recently in Hungary, at Vertészöllös.

Homo erectus did not stick to the crude choppers and other pebble tools which had characterized his first half-million years or so, without much change from the earlier tool-kit of the australopithecines. Around half a million years ago, starting perhaps in Africa, he began to make fairly large stone implements with more or less pointed ends, still quite crudely flaked, known as "hand axes," or, in French, "coups de poing" (Bordes 1968). Just how they were used is not clear, but they could have served to butcher large animals. Very slowly, during the next few hundred thousand years, they were improved in the degree of finish, and the flaking came to be controlled and made with a secondary tool rather than just by whacking off flakes with another stone. Meanwhile, more tool-types appeared: stone cleavers and crude scrapers. None of these tools seem to have had handles. Presumably wooden tools were also fashioned such as spears and clubs, and there may also have been wood or bark containers, along with straps and cords.

What caused the larger brains of *Homo erectus*? If, as some recent experiments suggest (they have been carried out so far only with rats), enriched experience produces increased brain growth in the individual, it is conceivable that generations of enriched experience and slightly enlarged brains might have conferred a selective advantage on individuals who had inherited genes making for slightly bigger skulls. Part of the enrichment may have been due to the greater complexities of tool-making and tool-using, and possibly to the more organized, carefully planned kind of hunting activity. Control of fire would have provided still another dimension to human culture, with its possibilities for cooking, light during the night, and possible uses in hunting drives. It is also usually supposed that language-using would create a need for more memory-storage and fine controls either of the vocal system, or perhaps only of the hands and fingers in gesture. The present marked verbal or language superiority of females has been attributed to the fact that in the course of language evolution, children have learned language from their mothers, and that it would have been more disadvantageous for females to suffer from various speech defects, than for males. Research in psycholinguistics and neurology indicates that the human brain must have had to add some highly complex functions in order to be able to decode swiftly spoken speech sounds. An interesting aspect of all this is that language, whether spoken or gestural, seems bound up with cerebral lateralization. This means the localization of certain functions on one of the two brain hemispheres, along with handedness. Not only are most human beings right-handed, but in the overwhelming number of cases, they are left-hemisphered for control of right hand, and left-hemisphered for language. The right half of the cerebral cortex appears to do better in the decoding of environmental noises, certain spatial problems, and, since it has come on to the cultural scene, music.

It is tempting to think that preferential handedness connected with tool-making and tool-using paved the way for the lateralization of gestural language and, still later, favored the same (left) hemisphere for the localization of speech control and decoding of speech sounds. By the time of Steinheim man, a fossil skull found in Germany, dating back to about 300,000 years, articulate speech may have been physically possible, according to the same research team which reported that *Homo erectus* probably did not have the capacity to make articulate speech sounds any more than the australopithecines.

The rates of evolutionary change do not seem to have been uniform over the vast range of *Homo erectus*. In far southern Africa, there are indications that a good many of the genes of *Homo erectus* survived until 30,000 years ago in the rugged, heavy-browed Rhodesian man (found in what is now Zambia), and an only slightly earlier find from Saldanha Bay, north of Cape Town. More unexpectedly, recent finds in southeastern Australia (at Kow Swamp) indicate that a similar belated *Homo erectus* strain may have survived there until 20,000 years or so ago. In the Australian case, however, it is not assumed that *Homo erectus* had reached that continent far back in the Pleistocene, but rather that a population carrying many *Homo erectus* genes had migrated eastward from what is now Indonesia, across the water-gaps dividing the Sunda from the New Guinea-Australian continental shelf toward the end of the Pleistocene, perhaps no more than 10,000 years before the remains were deposited at Kow Swamp.

The hand ax tool-tradition did not spread into Southeast Asia or China, and so was not coterminous with the distribution of *Homo erectus*. By 100,000 years ago, there seems to have been considerable diversity in the Old World not only in hominid physical types, but also in tool-kits. This tendency was to be accentuated over the next several tens of thousands of years.

Neanderthal Man and Other Early Forms of Modern Man

A gradual reduction in the cranial and facial features of *Homo erectus* began as far back as 300,000 years ago, in forms like Steinheim and Swanscombe man, from Germany and southern England. In a broad zone of Eurasia, North Africa, and the Middle East, there developed after 100,000 years or so ago a series of populations, probably interrelated and forming something like a "race," which we call *Neanderthal* after the first recognized discovery of this form in the Rhineland in 1856 (Coon 1963; Day 1965; Brace et al. 1971). These men were heavy-browed, long-headed, heavy-jawed, but with brains the same size as, or larger than, those of modern men. Their implements exhibit a definite advance in stone-working skill, mostly in the tool-tradition called "Mousterian" from a type-find site in southwestern France. Although they did not produce cave paintings or bone and antler tools of the kinds which became characteristic of the last half of the last glacial advance, they did bury their dead with offerings which at one cave site consisted of wildflowers, and some seem to have had some other ritualized activities having to do with disposal of cave-bear skulls. Over much of Eurasia, the climate of much of the Neanderthal or Mousterian time span was harsh, with great southward extensions of ice sheets, bordered by arctic tundra. Animal and plant life now restricted to the Arctic existed much farther to the south (Butzer 1964; Cornwall 1968).

Human forms somewhat resembling Neanderthal man also existed at this period in China. Other populations, presumably with persisting *Homo erectus* features, as noted, lived in southern Africa and the former peninsular landmass that is now Indonesia.

Some recent experimental work has indicated that Neanderthal man could have had an articulate spoken language, although his vocal quality would have sounded strange to our ears. The prevalence of flint scraping tools in the Mousterian tool-kit suggests that hide-working was well developed, and

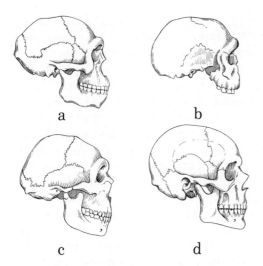

Figure 4. Lateral views of skulls: a. *Homo erectus,* near Peking, N. China; b. Steinheim man, W. Germany; c. *Homo sapiens,* Neanderthal man, Shanidar Cave, N. Iraq; d. *Homo sapiens sapiens,* modern man.

it is highly probable that Neanderthal man used animal skins or furs for protection from subarctic climate conditions.

Toward the midpoint in the final Wurm glacial period, 40,000 or so years ago, new features begin to appear, heralding the emergence of fully modern man—*Homo sapiens sapiens.* In particular, these indications show up in remains of cave-dwellers on Mount Carmel, Israel, where individuals seem to have combined some old Neanderthal traits with features which either suggest a rather rapid local evolution toward fully modern man, or intermixture with a modern form of mankind from some other region. If the latter is the explanation, we have not yet located the breeding ground for these essentially modern men. At this writing, the oldest Carbon[14] dated remains of modern man, about 43,000 years ago, come not from anywhere near the Middle East, but from a cave on the island of Borneo, in northern Sarawak (Niah Cave), where a youth was buried whose bones resemble those of some living people of Australia. Immense areas of southern Eurasia remain unexplored for ancient human remains, so it is conceivable that modern man may have evolved on the Indian subcontinent; another likely region still little known could have been southwestern Asia or northeastern Africa.

What were the specific traits of the "essentially modern-type man" who appeared at this time (Coon 1963, 1965)? Mainly they were cranial and facial features: smaller, weaker-looking jaws, with a tendency for a sharply pronounced chin rather than the blunt, rounded front of the lower jaw in previous forms, including Neanderthal Man. The whole midfacial region tends to lose its massive boniness, and in the area below the eyes and above the canine teeth a fairly deep depression becomes common: the "canine fossa." The nasal aperture generally gets narrower (although this is a fairly variable modern racial feature), and the protrusion of the fleshy and cartilaginous part of the nose above the face is more pronounced. Tooth size, especially of the molars, dwindles, though here again some living racial groups retain very large teeth. The forehead becomes fuller and rounder, and eyebrow ridges also diminish in prominence. Thickness of the skull bones also shows a marked decline. There seem to be fewer changes in the body below the neck. Though the skulls of most modern forms of mankind appear lighter and more flimsily constructed than those of the preceding groups we have been discussing, actual brain size did not exhibit any sharp upturn at this point. The fine, internal structure of the brain, especially of parts of the cerebral cortex, may have been developing capacities for more expert handling of language symbols, or quicker problem solving, for all we know, but the bony remains of the braincase do not tell us about this. Our only guide to possible (and by no means really certain) evolutionary improvements in the human brain over the past thirty to forty thousand years lies in the cultural record: the tools and weapons, artwork, indications of ritual, and, for the last five thousand years, the growth of civilization (Bordes 1968; Clark, G., 1967, 1969, 1970). While we may be pleased to assume that man's intellectual, aesthetic, and social capacities have been steadily improving over this time, it could be that all this progress, if we so choose to regard it, rests solely on accumulated learning, and that the people of 40,000 years ago would be, if alive today, just as capable of carrying on and developing modern civilization as people alive now.

Some biological changes have apparently taken place over the past 40,000 years in modern man. Tooth dimensions have been reduced, except in a few outlying populations. Reduction in tooth size may be related at least partly to the advent of agricultural foods and food processing such as cooking and milling. The consequences of life in more populous villages, rather than temporary camps, must have imposed a number of biological pressures (Struever 1971; Ucko and Dimbleby 1969). Still more recently, life in towns and cities has brought about further biological responses. To a considerable extent these new pressures appeared in the form of diseases formerly rare, including both contagious ones exacerbated by closer living conditions and poor sanitation, and chronic malnutrition arising from dependence on one or a few dietary staples instead of a diversified and more nutritious diet. Until quite recent times, practically until the nineteenth century, man paid a heavy price in illness for the advantages of living in large, more permanent communities.

Among the most important changes which may have mainly taken place since the advent of modern man are probably those we call "racial features"

(Coon 1965). The fossil bone and tooth remains do not inform us about the skin color, hair form, and similar traits of the people of the past. Only those racial attributes which are determinable from the skull and skeleton can be traced very far back into prehistory. It is quite clear from ancient paintings, however, that at least some of the major racial traits of mankind were present at least four to five thousand years ago. The ancient Egyptians depicted blacks with woolly hair and other recognizable racial features and also some less striking differences between themselves and the peoples of Libya to the west and the southwestern Asian area northeast of Egypt. Less clear-cut traces of the main racial groups of *Homo sapiens sapiens*, based on cranial and facial data, can be discerned in fossil human remains back to around 40,000 B.C. The first modern human skeleton found, from Niah Cave, seems to have been more or less "Australoid" in race. Next we have some probably Caucasoid individuals from Upper Paleolithic Europe, and Mongoloids from North China at the same period. In southern Africa are remains which show a resemblance to the modern Bushman and Hottentot. The oldest Negroid remains go back only about 10,000 years, from a find in Nigeria, West Africa. All of this suggests that the existing geographic races are all relatively recent manifestations, and that racialization is essentially a phenomenon of modern man. But the topic of racial origins, although a continuation of man's biological evolutionary history, really lies beyond the scope of this account.

We cannot really come to conclusions about an unfinished story. Sol Tax, a cultural anthropologist at the University of Chicago, expresses very well the continued role of evolution, both biological and cultural, in the case of mankind.

1. Culture is part of the biology of man, *of course, even though it is passed on socially and not through the genes. It is a characteristic of our species, as characteristic as the long neck is of the giraffe. The general biological questions asked about the giraffe's neck are also questions to be asked about the civilizations of man.*

2. Culture is part of the evolution of man. *Man is evolving continually as a species, perhaps more rapidly now than any other species. Hence, processes of natural selection and the like are presumably operative, but they are operative on the species, not on the particular cultures of communities of men.*

3. The term "evolution" is applied to both socially transmitted culture and gene-transmitted biology *because neither can establish an exclusive claim. However, there is no identity between the two usages. The cultural processes of continuity and change are different, and it is only by analogy, if at all, that one can speak of "natural selection," for example, in the development of cultures.*

Culture must be studied as part of the evolution of man; but culture change and growth must be studied in their own terms. Therefore, anthropologists legitimately study culture apart from the organisms who carry it.

4. Cultural behavior has a quality of arbitrariness, *because it does not flow through the genes and is therefore not anchored to the individual. This is seen most clearly in the arbitrariness of the symbols in language. Characteristically, therefore, cultures differ widely from community to community; the communities of men have this quality in common: each has its own special language, value systems, social systems, etc.*

5. The study of man becomes a comparative study of cultural difference within "genetic" sameness. *The species is uniform, with whatever individual differences there are in any population; but every community has its stamp. Both factors must be brought into the comparison.*[11]

We may question two of Tax's points, however: where he asserts that biological evolution operates only on our species as a whole, and not on particular cultural communities (No. 2, above), and No. 3, where he states that anthropologists legitimately only study culture apart from the organisms who carry it. Microevolution can be shown to operate even in very small human populations, such as closely inbred religious or ethnic groups, and cultural practices can, if sufficiently longlasting, enter the biological realm as factors in the selective process. If culture and biology have always been separate domains, we have no way to explain, for example, how man's language capacity has been perfected, or how a feedback process has apparently operated between man and his tools (such as food-grinding tools and their apparent effect on the dentition, in the very long run).

[11]Sol Tax. 1960. "The celebration, a personal view." In Sol Tax, ed., *Issues in evolution*. Vol. 3 of *Evolution after Darwin*. Chicago : University of Chicago Press, pp. 271-82.

What Comes Next?

Mankind is still evolving biologically, even under the conditions of modern civilization. With regard to the racial differences within the human species, it is possible that in the last few centuries a long-term tendency toward racial differentiation has been reversed and that the future will witness considerable racial blending and homogenization. The former local geographical pressures have been rendered less and less influential owing to the developments of modern technology. Just as cultural equipment has largely erased whatever biological advantage Eskimos or other Arctic peoples may have enjoyed as long-time inhabitants of that part of the earth, so mosquito-eradication programs and antimalarial drugs have made the protection against malaria afforded by inheritance of "half a dose" of sickle-cell anemia much less desirable. The invention of sunlight-reflecting clothing has long since offset the advantage of having a very darkly pigmented skin, just as the invention of artificial ways to bleach one's hair has practically destroyed the need, whatever its basis, to inherit blondism. However, the advantages or disadvantages of this or that inborn racial feature have by no means all been put aside by technology or changes in social attitudes. It may be a long time before the present racial variations in mankind actually vanish, or can be offset by cosmetic procedures, or simply cease to matter.

The foregoing account is a composite of some quite solid scientific evidence, plus speculation or guesswork, and the present writer's unbridled imagination. Most scientific subjects are really in the same boat, with numerous unanswered questions. In twenty or thirty years, portions of the foregoing story will seem very old-fashioned if not absurd. Other parts may be acceptable, but by then commonplace knowledge instead of exciting, and perhaps a small part of what has been said without solid backing at the present time will appear to have been remarkably insightful.

References and Suggested Readings

EVOLUTION BEFORE DARWIN

Chambers, Robert. 1969. *Vestiges of the natural history of creation.* Facsimile reprint, The Victorian Library. New York, Leicester: Humanities Press and Leicester University Press (with a new introduction by Gavin de Beer). First published anonymously in 1844.

Edwards, W. N. 1967. *The early history of paleontology.* London: Trustees of the British Museum (Natural History).

Gillispie, Charles Coulston. 1951. *Genesis and geology: a study in the relations of scientific thought, natural theology, and social opinion in Great Britain, 1790-1850.* Harvard Historical Studies, vol. 58. (Reprinted 1959, New York: Harper Torchbooks, Harper and Brothers, Publishers.)

Glass, Bentley; Temkin, Owsei; and Straus, William L., Jr., eds. 1959. *Forerunners of Darwin, 1745-1859.* Baltimore: Johns Hopkins Press (paperback edition, 1968).

Greene, John C. 1959. *The death of Adam.* Ames, Iowa: Iowa State University Press. (New York: New American Library edition, 1961.)

Lovejoy, Arthur O. 1936. *The great chain of being: a study of the history of an idea.* Cambridge, Mass.: Harvard University Press. (1960 paperback edition, New York: Harper and Row, Publishers.)

THE DARWINIAN REVOLUTION

Darwin, Charles. 1859. *Origin of species.* London (many later editions; 6th ed., 1872).

———. 1871. *The descent of man.* London (many later editions; 2nd ed., 1874).

———. 1872. *The expression of the emotions in man and animals.* London.

de Beer, Gavin. 1965. *Charles Darwin, a scientific biography.* New York: American Museum of Natural History and Doubleday Anchor Books.

Huxley, Thomas Henry. 1863. *Man's place in nature and a supplementary essay.* London: Watts and Company (reprinted 1921).

HUMAN ORIGINS–EARLIER VIEWS, 1900-1950

Hooton, Earnest A. 1946. *Up from the ape.* 2nd ed., rev. New York: Macmillan Co.

Osborn, Henry Fairfield. 1915. *Men of the Old Stone Age: their environment, life, and art.* New York: Charles Scribner's Sons (3rd ed., rev., 1922).

EVOLUTION, GENERAL

Alland, Alexander. 1969. *Evolution and human behaviour.* London: Tavistock Publications.

Cain, A. J. 1960. *Animal species and their evolution.* New York: Harper and Brothers, Harper Torchbooks. (1st ed., London, 1954.)

Dobzhansky, Theodosius. 1955. *Evolution, genetics and man.* New York: John Wiley and Sons.

Dunn, L. C. 1969. *Heredity and evolution in human populations.* Rev. ed. New York: Atheneum Publishers.

Kelly, Peter. 1966. *Evolution and its implications.* New York: Hawthorn Books.

Tax, Sol, ed. 1960. *The evolution of man: mind, culture, and society.* Vol. 2 of *Evolution after Darwin.* Chicago: University of Chicago Press.

VERTEBRATE EVOLUTION

Romer, A. S. 1971. *The vertebrate story.* Chicago: University of Chicago Press. (Originally published in two volumes under the title *Man and the vertebrates*, 1954.)

Time-Life Books, Editors. 1972. *Life before man.* The Emergence of Man. New York: Time-Life Books.

PRIMATE EVOLUTION

Clark, W. E. Le Gros, 1959. *The antecedents of man: an introduction to the evolution of the primates.* New York and Evanston: Harper and Row, Publishers.

Pilbeam, David. 1972. *The ascent of man: an introduction to human evolution.* New York: Macmillan Co.

Simons, Elwyn L. 1972. *Primate evolution: an introduction to man's place in nature.* New York: Macmillan Co.

PRIMATE BEHAVIOR

Dolhinow, Phyllis, ed. 1972. *Primate patterns.* New York: Holt, Rinehart and Winston.

Eimerl, Sarel, and DeVore, Irven. 1965. *The primates.* New York: Time Incorporated.

Goodall, Jane van Lawick . 1971. *In the shadow of man*. Boston: Houghton Mifflin Co.
Jolly, Alison. 1972. *The evolution of primate behavior*. New York: Macmillan Co.
Kummer, Hans. 1971. *Primate societies: group techniques of ecological adaptation*. Chicago and New York: Aldine-Atherton.
Morris, Desmond, ed. 1967. *Primate ethology*. Chicago: Aldine Publishing Co.
Quiatt, Duane D., ed. 1972. *Primates on primates: approaches to the analysis of nonhuman primate social behavior*. Minneapolis: Burgess Publishing Co.

HUMAN EVOLUTION, GENERAL
Bishop, Walter W., and Clark, J. Desmond, eds. 1967. *Background to evolution in Africa*. Chicago: University of Chicago Press.
Howell, F. Clark, and Bourlière, François, eds. 1963. *African ecology and human evolution*. Viking Fund Publications in Anthropology, No. 36. New York: Wenner-Gren.
Washburn, Sherwood L., ed. 1963. *Classification and human evolution*. Viking Fund Publications in Anthropology, No. 37. New York: Wenner-Gren.

FOSSIL MAN
Brace, C. Loring, and Montagu, M. F. Ashley. 1965. *Man's evolution: an introduction to physical anthropology*. New York: Macmillan Co.
Brace, C. Loring, Nelson, Harry, and Korn, Noel. 1971. *Atlas of fossil man*. New York: Holt, Rinehart and Winston.
Campbell, Bernard. 1966. *Human evolution: an introduction to man's adaptation*. Chicago: Aldine Publishing Co.
Coon, Carleton S. 1963. *The origin of races*. New York: Alfred A. Knopf.
Dart, Raymond A. (with Dennis Craig). 1959. *Adventures with the missing link*. New York: Harper and Brothers, Publishers.
Day, Michael. 1965. *Guide to fossil man: a handbook of human paleontology*. Cleveland and New York: Meridian Books, World Publishing Co.
Edey, Maitland, and the Editors of Time-Life Books. 1972. *The missing link*. New York: Time-Life Books.
Howell, F. Clark, and the Editors of *Life*. 1965. *Early man*. New York: Time-Life Books.
Leakey, Louis S. B., and Goodall, Vanne Morris. 1969. *Unveiling man's origins*. Cambridge, Mass.: Schenkman Publishing Co.

Leakey, Richard E. F. 1973. "Evidence for an advanced Plio-Pleistocene Hominid from East Rudolf, Kenya." *Nature* vol. 242 (April 13): 447-450.

Napier, John. 1970. *The roots of mankind: the story of man and his ancestors.* London: George Allen and Unwin.

Pfeiffer, John E. 1972. *The emergence of man.* 2nd ed. rev., New York: Harper and Row, Publishers.

Young, Louise B., ed. 1970. *Evolution of man.* New York: Oxford University Press.

EVOLUTION OF HOMINID BEHAVIOR

Ardrey, Robert. 1961. *African genesis.* New York: Atheneum Publishers. (Paperback edition, New York: Dell Publishing Co., 1963.)

_____. 1968. *The territorial imperative: a personal inquiry into the animal origins of property and nations.* New York: Dell Publishing Co.

Hass, Hans. 1970. *The human animal: the mystery of man's behavior.* Translated by J. Maxwell Brownjohn. London: Hodder and Stoughton.

Hofer, H., and Altner, G. 1972. *Die Sonderstellung des Menschen.* Stuttgart: Gustav Fischer Verlag.

Lorenz, Konrad. 1967. *On aggression.* Translated by Marjorie Kerr Wilson. New York: Bantam Books, Harcourt, Brace and World.

Montagu, M. F. Ashley, ed. 1965. *The human revolution.* Cleveland: World Publishing Co.

_____. 1968. *Culture: man's adaptive dimension.* London and Oxford: Oxford University Press.

Morris, Desmond. 1967. *The naked ape: a zoologist's study of the human animal.* New York: McGraw-Hill Book Co.

Spuhler, J. N., ed. 1959. *The evolution of man's capacity for culture.* Detroit: Wayne State University Press. (Paperback edition, 1965.)

Washburn, Sherwood L., ed. 1961. *Social life of early man.* Viking Fund Publications in Anthropology, No. 31. New York: Wenner-Gren.

ARCHAEOLOGICAL EVIDENCE OF EARLY HUMAN BEHAVIOR

Bordes, François. 1968. *The Old Stone Age.* Translated by J. E. Anderson. London: Weidenfeld and Nicolson.

Butzer, Karl W. 1964. *Environment and archaeology: an introduction to Pleistocene geography.* Chicago: Aldine Publishing Co.

Clark, Grahame. 1969. *World prehistory—an outline.* Rev. ed. Cambridge: Cambridge University Press. (1st ed. 1961.)

_____. 1967. *The Stone Age hunters.* New York: McGraw-Hill Book Co.

_____. 1970. *Aspects of prehistory*. Berkeley: University of California Press.

Cornwall, I. W. 1968. *Prehistoric animals and their hunters*. London: Faber and Faber.

Lee, Richard B., and Devore, Irven, eds. 1968. *Man the hunter*. Chicago: Aldine Publishing Co.

Struever, Stuart, ed. 1971. *Prehistoric agriculture*. American Museum Source Books in Anthropology. Garden City, N.Y.: Natural History Press.

Ucko, Peter J., and Dimbleby, G. W., eds. 1969. *The domestication and exploitation of plants and animals*. Chicago: Aldine Publishing Co.

ORIGINS OF THE MODERN RACES OF MAN

Coon, Carleton S. 1965. *The living races of man*. New York: Alfred A. Knopf.

GENERAL WORKS ON PHYSICAL ANTHROPOLOGY

Bleibtreu, Hermann K., and Downs, James F., eds. 1971. *Human variation: readings in physical anthropology*. Beverly Hills, Calif.: Glencoe Press, Macmillan Co.

Buettner-Janusch, John. 1966. *Origins of man: physical anthropology*. New York: John Wiley and Sons.

Cohen, Yehudi, ed. 1968. *Man in adaptation: the biosocial background*. Chicago: Aldine Publishing Co.

Dolhinow, Phyllis, and Sarich, Vincent M., eds. 1971. *Background for man: readings in physical anthropology*. Boston: Little, Brown and Co.

Harrison, Richard J., and Montagna, William. 1969. *Man*. New York: Appleton-Century-Crofts.

Kelso, A. J. 1970. *Physical anthropology, an introduction*. Philadelphia: Lippincott.

Young, J. Z. 1971. *An introduction to the study of man*. Oxford: Oxford University Press.